RESTORING AMERICA

A VOLUME IN THE SERIES
PUBLIC HISTORY IN HISTORICAL PERSPECTIVE
Edited by
Marla R. Miller

RESTORING AMERICA

HISTORIC PRESERVATION
and the
NEW DEAL

Stephanie Gray

University of Massachusetts Press

AMHERST AND BOSTON

ISBN 978-1-62534-897-5 (paper); 898-2 (hardcover)

Designed by Jen Jackowitz
Set in Adobe Caslon Pro
Printed and bound by Books International, Inc.

Cover design by adam b. bohannon
Cover photo by unknown photographer, *NYA workers construct kiln for ceramic studio*, 1939.
Courtesy of University of Texas at San Antonio Historical Archives

Library of Congress Cataloging-in-Publication Data
A catalog record for this book is available from the Library of Congress.

British Library Cataloguing-in-Publication Data
A catalog record for this book is available from the British Library.

The authorized representative in the EU for product safety and compliance is
Mare-Nostrum Group.
Email: gpsr@mare-nostrum.co.uk
Physical address: Mare-Nostrum Group B.V., Mauritskade 21D,
1091 GC Amsterdam, The Netherlands

For my mother, Lucy Gray

.

Contents

CONCLUSION

192

List of Illustrations

Acknowledgments

This book is a culmination of hard work, persistence, an abiding love for history and old buildings, and generous support from family, friends, and institutions. As is the case with all professional and personal accomplishments, the joy of completion is accompanied by gratitude for the people and places who helped make the writing and publication of this book possible.

First, I count myself lucky to work with the folks at the University of Massachusetts Press to whom I am profoundly grateful for shepherding me through publication of this monograph. From the beginning, editor-in-chief Matt Becker has demystified the process, answered all of my questions, and helped keep me moving toward the finish line. Thank you to the production and marketing team—Ben Kimball, Sally Nichols, and Chelsey Harris—for polishing and promoting this work, and to copy editor Margaret A. Hogan for her thoughtful and thorough review. To the two anonymous reviewers who read the manuscript, thank you for the productive feedback and encouraging remarks that guided me while refining and reframing the book's argument. The resulting work is stronger for your insights.

Two universities have supported this project from start to finish. The first is the University of South Carolina, where this book began as a research paper in Marjorie Spruill's graduate seminar. I am especially grateful to the members of my dissertation committee whose critical feedback helped transform dissertation into book: Lauren Sklaroff, Robert Weyeneth, Lydia Mattice Brandt, and Patricia Sullivan. As my doctoral advisor, Lauren gave me language to describe this project's significance within American cultural history and demonstrated admirable balance between personal and professional life. Bob, paragon of excellence in the field of public history, has offered unwavering support since day one and continues to be a fountain of knowledge and wisdom. I hope I have earned the title of "esteemed alum." Lydia puts practice into action as both a serious scholar and a community-engaged preservationist. I am grateful for her invaluable mentorship and friendship during my years in Columbia. I am also thankful to the Department of History and Graduate School of the University of South Carolina for providing essential funding through travel awards, the SPARC Graduate Research Grant Program, and the Bilinski Fellowship.

The second institution that helped fuel this book is Duquesne University, which has been a wonderful home for me to grow as a scholar, teacher, and public historian. My first thank you belongs to John C. Mitcham, the chair of the Department of History and my mentor. John read almost the entire manuscript and provided crucial feedback from outside of the public history field. Beyond that, he invariably steered me in the right direction while navigating academic life. I did not expect to find the kind of friend and mentor I have in him, and I am deeply grateful. I am also very fortunate to have Jennifer Whitmer Taylor as my teammate in building Duquesne's Public History Program. Her energy, commitment to putting history to work for good, and dedication to helping students become the best public historians they can be is inspiring. While deep in the trenches of her own manuscript, she unfailingly championed my work and sometimes took on more tasks for herself to protect my time. I could not have asked for a better collaborator and friend in the department. I would also like to thank our department administrative assistant Sam Hicks-Gandy for handling payments and permissions and for offering moral support. Her friendship has made the toughest days of writing easier. To Dean Kristine Blair of the McAnulty College and Graduate School of Liberal Arts, thank you for your support, especially of junior faculty, and for modeling excellent leadership. Many thanks to Julie Dougherty and Janet Sculimbrene in the Dean's Office for assisting with all of the administrative work associated with the Wimmer Family Foundation, which provided critical research funds for this project. Finally, I am grateful to the graduate students of the Department of History whose contributions to class discussions helped shape the way I think about the field and my work. In particular, I want to express my appreciation to my graduate assistants Bryton Altenbach,

Hannah LeComte, and Glenna Van Dyke for their patience, diligence, and hours spent reading drafts, transcribing sources, and communicating with archives for me.

That leads me to the fact that the historian's work is only possible because of practitioners. I am grateful for the assistance of the archivists, librarians, aides, graduate students, and other staff at the following repositories: the Charles Lindbergh House and Museum, Connecticut Museum of Culture and History, Connecticut State Library, Dolph Briscoe Center for American History at the University of Texas at Austin, Henry Whitfield State Museum, Minnesota Historical Society, National Archives at College Park, San Antonio Conservation Society, San Antonio Public Library, South Carolina Historical Society, South Caroliniana Library at the University of South Carolina, Sterling Memorial Library at Yale University, and University of Texas at San Antonio Special Collections.

Working at the James A. Garfield National Historic Site throughout college summers showed me what "history put to work in the world" really means. I thank the wonderful volunteers and staff, especially my supervisors Todd Arrington and Mary Lintern, for teaching me how to do public history on the frontline at an undeservedly underrated president's family home. And to my favorite history teachers: Ric Doringo, Jeremy King, and Desmond Fitz-Gibbon, thank you for making history compelling and instructive, and for showing me that studying history can lead to a fulfilling, enjoyable, and important career.

Beyond professional life, I have a terrific group of people in my circle who sustain me with their generosity, humor, intellectual curiosity, and camaraderie. To my longtime friends outside of academia—Jordan Arnold, Michael Cabrelli, Cassie Denbow, Cindy Heiple, Soo Jin Lee, Jamie Lovett, Jill Wollenberg Meyers, Abi Poe, Elizabeth Stiles, and Abby Curtin Teare—your steadfast friendship has provided a stable anchor in my life for decades. The friends who experienced the joys and challenges of graduate school alongside me deserve special recognition as well: Janie Campbell, Alyssa Constad, Kristie DaFoe, Diana Olding, Kayla Boyer Halberg, Samuel King, Robert Olguin, Andrew Walgren, and Virginia Harness. Lastly, to my Pittsburgh crew, especially Jeff and Kathy Lambert, Erin Conlin, Joe Burke, and Stephen Newmyer, it has been such a delight to build a community with you all. A special thanks to Arthur and Lauren Sugden for their warm friendship and the many laughs we have shared over games, meals, and holidays.

To David Gray Widder: I feel profoundly lucky that you entered my life when you did. Thank you for your support, understanding, and excitement when I needed it most. To the entire Widder family with whom I spent time while working on this book: thank you for the much-needed cheer and laughter.

The final thank you belongs to my family, who has shown endless and unquestioning confidence that I was always doing what I needed to do. To my father, Jim Gray; sister, Victoria Gray; grandma, Emma Simoncic; and aunt, Kathy Livingston, thank you for supporting me always. Lastly, I thank my mom, Lucy Gray, to whom this book is dedicated. It was your love of reading and interest in history that got me started on this path when I was very young. You always encouraged me to read, to learn, to ask questions, and to use education as a tool of empowerment. This book is the result of your love and what you have taught me to love.

RESTORING AMERICA

Introduction

Scattered across the broad face of our land are hundreds of historic places or structures—milestones in the development of a nation. Many an American community, busy with the problems of today and tomorrow, had put off the maintenance, repair or improvement of a cherished historic shrine until it became virtually a ruin. But when local officials found their own jobless workers available, at WPA pay, they hastened to provide materials and start the rehabilitation of old forts, old homes and other sites at which significant bits of history once were enacted. Often an historical association or some other agency had begun to restore a shrine and then had been unable to complete the work. Without federal aid the shrine would never have been preserved for posterity.

Historic Shrine Summary, April 18, 1939,
Works Progress Administration

In February 1937, all state administrators of the Works Progress Administration (WPA), a New Deal federal work relief program, received a letter from the agency's Division of Information in Washington, DC. The letter reported that the WPA's central office was preparing a catalogue of "outstanding projects through which historic shrines such as historical parks, forts, residences, missions and other shrines have been restored or preserved." It asked the state administrators to send a narrative account, photographs, and a brief statement of historical significance about ongoing projects "of some national interest" that could be considered "shrine" restorations. As guidance, the letter provided specific examples of the type of work the Division of Information desired to feature in the catalogue: it named as model sites the Dock Street Theatre in Charleston, South Carolina; Lincoln Pioneer Village in Rockport, Indiana; and Fort Jefferson in the Florida Keys' Dry Tortugas islands, among others.[1] Additional work to be featured included "such manual projects as the restoration of the Lost Colony settlement on Roanoke Island, the repairing of the Statue of Liberty, the reconstruction of Civil War Fort Negley in Tennessee, [and] the digging of artifacts in Southwestern pueblos." Provisionally titled "America Down the Years: Preserving the Past for Posterity," the booklet would draw further attention to "other WPA work of historic American interest," citing contributions of the Historical Records Survey, the Historic American Buildings Survey, and the Merchant Marine Survey, other New Deal initiatives that recorded significant historic structures and vessels.[2]

Although the WPA "considered very important" the publication of this illustrated booklet, the agency's central office never did produce a special catalogue solely devoted to historic American shrines.[3] Rather, the restoration projects appeared in a 1938 inventory of the impressive work the WPA completed during the relief program's first two years of operation from 1935 to 1937. The section titled "Historic Shrines" included a narrative overview of the agency's preservation activity and a two-page picture spread of exemplary sites from the nearly one hundred eligible projects nationwide either already completed or in operation (fig. 1). "Every period in America's history is represented, from the days of Indian supremacy to yesterday," the publication boasted.[4] The highlighted projects presented a diverse portrait of America's historic built environment, ranging in age, architectural style, and ethnic and racial communities of origin. From Spanish missions in the South and West to Revolutionary War battlefields on the East Coast, from birthplaces and family homes of presidents and other famous men to utilitarian military forts scattered across the country, the restored sites materialized in physical form foundational and mythical narratives of American history. In attending to historic places as New Deal projects, the booklet asserted, WPA workers helped preserve for posterity renowned and revered national and local "cradles of democracy" that served as "landmarks along the route of American development."[5]

This New Deal preservation work reflected both pragmatic and ideological aspirations: it repaired and improved neglected sites of history but also, importantly, provided redress for an unemployed and culturally unmoored nation. A WPA press release explained that the "re-creation of these shrines has served a double purpose—employment for the deserving needy, and a visible reminder of the hardships endured by the Nation's founders."[6] These projects offered, in other words, both recompense in the form of stable jobs and evidence that past generations had withstood turbulent times, a physical and emotional recalling of historic trials that tested Americans. The political and economic backdrop of this 1930s preservation work was, of course, the Great Depression, sparked by the Wall Street stock market crash of October 1929. Throughout the next decade, Americans suffered in the throes of the most formidable economic catastrophe they had ever experienced: a stagger-ing one-fifth to one-quarter of the nation's workers faced unemployment, breadlines wrapped around the block, schools closed when children became truants in order to help their families, and farmers watched as their crops withered away in failing markets. The financial, physical, and emotional hard-ships the Great Depression wrought on Americans appeared visually in the ubiquitous shantytowns, soup kitchens, and abandoned homes that dotted the national landscape. People everywhere grappled with the question of what it meant to be American in the face of immense economic, political, agricultural, and social upheavals. The Depression's challenge to the narrative of American prosperity and progress led to an existential crisis of national identity.

When Democrat Franklin Roosevelt stepped into office in early 1933, he immediately began pushing through legislation that promised to put the United States on the path of recovery. Created during both the First (1933) and Second (1935) New Deals, his so-called alphabet agencies attended to the financial doldrums, labor woes, and material scarcities of banks, businesses, families, and farms. In addition to offering economic and environmental relief, the New Deal administration created bureaucratic infrastructure that promoted cultural production and allowed for diverse artistic expressions to flourish. Throughout the 1930s and early 1940s, the WPA's Federal Project Number One set writers, artists, musicians, performers, and architects to the task of capturing America's historical and contemporary cultural scene in a variety of mediums. They recorded oral histories, published state guide-books, painted murals, and measured buildings to document the fractured yet indomitable "American way" of life during the Depression years.[7]

The agency's sponsorship of historic landmark restoration in this portfolio of political and cultural activity is lesser known. Repairing and improving

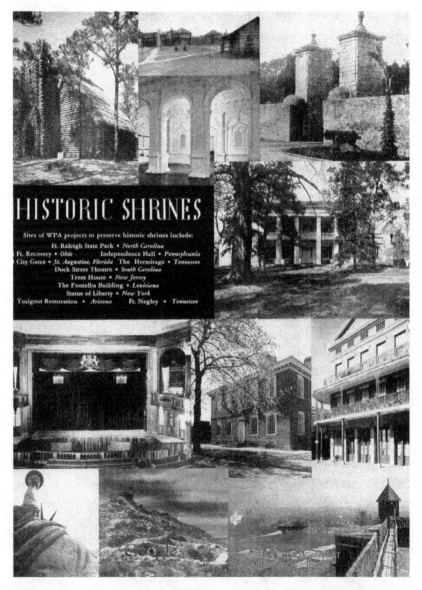

FIGURES IA–IB. Historic shrines pictured in the 1938 WPA inventory included Fort Raleigh State Park, North Carolina; Fort Recovery, Ohio; Independence Hall, Pennsylvania; City Gates, St. Augustine, Florida; The Hermitage, Tennessee; Dock Street Theatre, South Carolina; Trent House, New Jersey; The Pontalba Building, Louisiana; Statue of Liberty, New York; Tuzigoot Restoration, Arizona; Fort Negley, Tennessee; Flag House, Maryland; Old Court House, Delaware; Jumel Mansion, New York; The

Sites of WPA projects to preserve
historic shrines include:

Flag House • *Maryland*
Old Court House • *Delaware*
Jumel Mansion • *New York*
The Cabildo Fireplace • *Louisiana*
Ft. Niagara • *New York*
Fort Pike • *Louisiana*
Lincoln Village • *Indiana*
McDowell House • *Kentucky*
Faneuil Hall • *Massachusetts*
Fort Jefferson • *Florida*

Cabildo Fireplace, Louisiana; Fort Niagara, New York; Fort Pike, Louisiana; Lincoln Village, Indiana; McDowell House, Kentucky; Faneuil Hall, Massachusetts; and Fort Jefferson, Florida. Source: Works Progress Administration, *Inventory: An Appraisal of the Results of the Works Progress Administration* (Washington, DC: U.S. Government Printing Office, 1938), 33–34.

"cherished historic shrines" became an important method of working through the practical and philosophical cataclysms generated by the Depression.[8] Inspired by the Democratic administration's unprecedented support of the arts *and* its federal investment in public works projects, elected leaders, civic organizations, local boosters, and architects and artists skillfully leveraged New Deal funds to shape their local built environments through historic preservation efforts. Starting in 1935, the WPA and the National Youth Administration (NYA), which originated as a division of the WPA, began to finance preservation projects proposed by local and state sponsors that pragmatically put Americans to work, invested in the municipal development of historic infrastructure, and empowered local governments and civic-cultural elites to mythologize and capitalize on historical narratives about their communities.

This book explores *how* and *why* Americans employed historic preservation as a tool to address the political, economic, and social dilemmas they encountered during the Depression. Underwritten by the New Deal government, the projects' nonfederal sponsors—city or state government officials often encouraged by leading civic organizations—used restorations of popular historic sites to execute their own political agendas and make tangible statements about the cultural identity and economic future of a particular place. This productive federal-local partnership resulted from the "cooperative federalism" introduced by the New Deal. Before the 1930s, cities were considered wards of the state and rarely involved the federal government directly. State governments, too, spent their funds mostly on roads and schools, limiting their spending on direct relief or any kind of welfare program. But the Roosevelt administration restructured the relationship between federal and local levels whereby the U.S. government assumed greater responsibility for solving problems in the states and cities.[9] With more Democrats sitting in government offices and unprecedented availability of federal funds to draw on for direct unemployment relief, public works, agricultural reform, and more, there resulted in the 1930s "a perceptive growth in awareness of the need for positive government," argued historian James T. Patterson.[10]

Thus, New Deal cooperative federalism and an expanding outlook on positive government fundamentally changed the relationship between American municipal administrations and the national government. Mayors like Fiorello La Guardia of New York City eagerly capitalized on the newly accessible federal assistance to fund work relief, social services, public housing, and civic infrastructure projects like bridges, parks, and airports. Other city leaders and state governors came to realize that through grants-in-aid and work relief programs they could reap benefits from cooperative federalism without paying the bill.[11] As Robert D. Leighninger Jr., who has studied New Deal public infrastructure, explains, 1930s federalism "was truly cooperative, usually combining fiscal centralization with administrative decentralization."[12]

The New Deal chapter in the history of the American preservation movement, therefore, presents a model of *cooperative preservation* that resulted from the period's experiment in political federalism. Decentralized New Deal agencies like the WPA and NYA empowered nonfederal actors to direct historic preservation efforts and introduced a new critical player in their execution: the national government. In doing so, the preservation field during the Depression transitioned from primarily private activity that was either small and voluntary or large-scale and corporate, to undertakings paid for with federal funds, managed by local politicians and civic organizations, and designed to aid in economic recovery by boosting local tourism. The federal work relief programs gave the projects structure, oversight, and funding but extended autonomy to the local and state sponsors over historical, architectural, and promotional decisions. By employing architecture and material culture to reconstruct, recreate, and commercialize history, across the United States politicians, architects, craftspeople, and New Deal laborers elevated historic preservation as a powerful form of community revitalization. Therefore, while these preservation projects operated within the federal New Deal apparatus, the nonfederal sponsors extensively shaped the political contours of the projects. The federal government, hence, cooperated in the execution of community-based preservation work.

During the WPA's existence from 1935 to 1943, the agency ultimately funded nearly two hundred architectural restorations, in addition to archaeological projects and monument restorations, at the behest of local and state sponsors (see appendix A). The projects were located all across the nation, from West Coast to East Coast, although the largest concentration of sites was in the Northeast and Mid-Atlantic with the states of Massachusetts, New York, and Pennsylvania boasting the most shrine restoration projects. In addition to their scattered geography, the WPA's architectural and archaeological projects represented an expansive historical timeline, covering "prehistory"—by which the agency meant Indigenous history pre-European contact—to the twentieth century, with the majority of the historic sites dating to the eighteenth and nineteenth centuries. Thematically, the shrines convened around major nation-building events (like the American Revolution and the Civil War); influential men celebrated for their notable political, military, medical, or literary achievements (such as Andrew Jackson and Walt Whitman); and important civic institutions (including town halls and courthouses). They ranged in terms of their recognized level of significance, with some of the most famous buildings and structures of American history receiving maintenance, including Independence Hall in Philadelphia, Faneuil Hall in Boston, and the Statue of Liberty in New York City. Most shrines, however, enjoyed either local or regional acclaim rather than national, such as Mission San José in San Antonio, the Cabildo in New Orleans, and Tuzigoot Pueblo near Clarkdale, Arizona.

WPA literature about the shrines explained that the agency "extend[ed] a helping hand" to projects that without federal assistance could not be completed.[13] It funded the restoration of Grant's Tomb in New York City, for example, because "the depression caused it to be abandoned." The Grant Monument Association eagerly capitalized on the "opportunity for the Federal Works Progress Administration to provide funds and labor to do the necessary work."[14] The WPA acknowledged the role of local historical organizations in instigating worthwhile projects that helped depressed local economies, such as the Lincoln Pioneer Village in Rockport, Indiana, which was pushed forward by the Business Men's Association and the Spencer County Historical Society. In designating the reconstructed village where the future president lived between ages seven and twenty-one as one of the three best projects in the state, the WPA claimed, "It is not only because of the memory of Lincoln in this country that this project is favored, but its far reaching social and business effects mark it in numerous ways as proof of the fruits of WPA both present and future." Workers learned new building skills in reconstructing nineteenth-century log cabins, and the popular site honoring the Great Emancipator brought paying tourists to the city.[15] The agency lauded other projects for their role in creating a sense of community during the otherwise dismal period; the restoration of the Vernon-Wister House in Germantown, Philadelphia, proposed by the Germantown Community Council, turned the historic building "sliding so surely into decay through lack of attention . . . into a cultural community centre."[16] Considered collectively, these restored shrines played a critical role in reviving American social life with assistance from the federal government.

This book illustrates that historic preservation emerged as an important method of cultural production, an instrument of economic reconstruction, and a striking expression of political theater in the 1930s and 1940s. To make this case, it focuses on four major historic site restorations between the years 1935 and 1942: (1) the Dock Street Theatre in Charleston, South Carolina (1935–37); (2) the Henry Whitfield State Museum in Guilford, Connecticut (1935–37); (3) the Charles A. Lindbergh Boyhood Home and State Park in Little Falls, Minnesota (1937–41); and (4) the La Villita Historic Arts Village in San Antonio, Texas (1939–42). The WPA funded the first three restorations, while the NYA administered the fourth. From a popular colonial theater in the Deep South, to a Puritan minister's home in New England, to famed aviator Charles Lindbergh's modest farmhouse and lands of the Upper Midwest, to a multicultural Indigenous-European-Mexican arts village in the central South, the book's examples demonstrate the broad array of New Deal cooperative preservation work. As a collection, they represent geographic diversity, distinct architectural types and building ages, and differing amounts of financial assistance and attention from their funding agency's state and national offices. They take place in small towns (in Connecticut and Minnesota) and sizable metropolises (Charleston and San Antonio), and present local political climates that spanned the ideological spectrum, expressing varying levels of commitment to Roosevelt's New Deal liberalism.

Yet, the four case studies have in common important characteristics that make them appropriate, illuminating, and didactic sites for analysis. First, they all involved projects that left substantial archival records, both textual and visual. Second, all four examples are extant sites existing today mostly in the physical form they assumed during the 1930s and 1940s, which allowed me to visit them and spatially engage in similar ways to the historical actors described herein. Third, and critical to my argument, while all projects attracted a degree of attention from the state and central WPA or NYA offices, they were debated and propagandized most robustly at the local level. To that end, my goal is not to scrutinize the scale of the sponsoring federal agencies' interest in these particular projects but to uncover why local politicians and other influential community actors turned to historic preservation in a time of crisis. My focus here consequently is not on the WPA and NYA laborers themselves nor the bureaucratic administration of the two federal agencies encountered through the case studies; their mechanics as New Deal programs are addressed in chapter 1 but are covered in full detail in other scholarly works.

For each case study, I examine the localized context of preservation efforts by interrogating the political, economic, and cultural ambitions and anxieties of the people and groups with decision-making power, and the relationship between the federal program and local actors throughout the execution of this historical memory work. Across the four projects, we are introduced to a wide cast of characters, including enterprising politicians, renowned restoration architects, patriotic societies, leading regional preservation organizations, modern celebrities, professional artists, and New Deal officials. The maneuverings of these historical actors reveal several themes of New Deal cooperative preservation: the continued professionalization of the field with an expanded role for experts and masculinized leadership, race and class as powerful forces in the displacement of residents and transformation of blighted areas, and the commodification of history and memory in the development of local tourism industries. Moreover, this collection of projects allows me to put historic preservation into conversation with other political and cultural phenomena of the Depression era, from the modern celebrity craze, to Pan-American cooperation, to the colonial revival architectural fad, to increased enjoyment of America's outdoors as public recreation.

To understand how historical actors situated locally conceived and executed each case study in cooperation with their sponsoring New Deal agency, I draw on government records including official correspondence among federal, regional, state, and community figures; personal letters; memoranda; architectural notes and drawings; and New Deal publications and promotional materials. To further contextualize the local contours of each project's political and cultural landscape, I make heavy use of local newspapers, organizational and institutional records, personal memoirs, and oral histories. I also examine media such as advertisements, photographs, and tourism materials to understand the rhetoric surrounding the projects' design, construction, promotion, and reception.

In analyzing these projects as a cultural and public historian, I approach each of the historic sites as both a physical place and an idea of place; that is to say, I explore both the tangible foundations and intangible myths that are part of placemaking.[17] The built environment—a collective term to describe the combined landscape of architecture, material culture, and nature—affords an analytical window into motivations, cultural trends, aesthetic choices, and human desires that do not always appear in written forms. In his environmental history of the Civilian Conservation Corps, Neil M. Maher considers landscapes as cultural sources that can be "'read' as one would read court records, census tracts, and government documents for clues to the society that created them" and for "associated political transformations occurring within culture." I similarly consider cultural landscapes as more than "historical backdrop" or "contextual filler," and practice deep reading of the built environment to explore the powerful intersections among place, culture, politics, and power.[18]

Because local actors exerted a high degree of control over these preservation projects, I view the shrine restorations as manifestations of "vernacular memory" more than "official memory." As historian John Bodnar aptly explained, "Vernacular expressions convey what social reality feels like rather than what it should be like."[19] In their restored forms the sites intentionally made statements about community historical identity rather than collective national identity. Although tied together in operation under the New Deal work relief program, they teach us about grassroots commemoration—homegrown acts of remembering, memorializing, and politicizing the past. Therefore, while I have analyzed in depth four projects here, any one of the almost two hundred historic shrines sponsored by the WPA during its eight years could tell an instructive story about how local and state actors cooperated in preservation endeavors to advance political and social agendas. I hope that scholars will find value in examining other New Deal restorations not covered in this book.

A note on language in this study is necessary. Throughout, I primarily use the term "restoration" to describe the broad strokes of preservation work occurring in each case study for two reasons. First, it serves as a convenient umbrella term for all preservation activity, and second, it is the term historical actors themselves generally used to describe these New Deal projects. While we would engage more precise terminology today to explain the methodologies involved in each project, in the 1930s and 1940s preservation professionals, government officials, and the public used the word "restoration" to refer to a range of activity that indeed restored but also preserved, rehabilitated, and reconstructed historic sites.[20]

The second term that requires critical engagement is "shrine." Referring to historic sites as shrines is a legacy of nineteenth-century commemorative activity and historical tourism development, which I address in chapter 1. However, the use of the word also reflects the WPA's intentional positive framing of its preservation work; there was no concerted effort at the time to preserve what today we might call difficult, dark, or shameful chapters of American history. A clear example of the WPA shrines' limitations is the fact there was only one reference in the agency's 1938 inventory to the group of people who labored to construct many of these shrines: Fort Negley in Nashville, a Union fortification during the Civil War, was described as "built by impressed slaves."[21] Otherwise, acknowledgment of the enslaved labor involved in many of the shrines identified across the nation was generally missing. Historical narratives of forced assimilation, racial and gender discrimination, slavery and other forms of labor exploitation, and settler colonialism were likewise ignored. While Indigenous history featured prominently in the WPA's archaeology projects, their categorization as "prehistoric" perpetuated a romantic primitivism that positioned Native peoples as premodern others.[22]

It is important, then, to acknowledge that the WPA propagandized the shrine restorations because they were, by and large, uncontroversial political projects. Part of the reason for that was their general focus on white history, in terms of both *what* historical narratives were prioritized through the preservation of the built environment and *who* managed the restorations. Although the restorations represent a diverse and storied landscape of American history, there are obvious limitations to their democratic nature determined by who was empowered to shape what historic places—and therefore whose history—was included in the shrine restoration program and promoted in agency publications. While marginalized Americans found employment opportunities and created outlets for political and cultural expressions through various New Deal initiatives, the historic preservation projects covered here do not tap into the important facet of racial progressivism evident in some other federal endeavors of the period.[23]

This book conceives of historic preservation as both a cultural practice and a political process that ties directly into New Deal state-building, consolidation of control over place/space, and historical memory-making. Consequently, this interpretive work is informed by scholarship on the New Deal, historic preservation, public history, and American cultural history. First, this book joins the work of scholars who vigorously have looked to the cultural work of New Deal programs to understand the broader political, economic, and social implications of the Roosevelt administration's policies.[24] In particular, *Restoring America* contributes to the rich literature about the New Deal's search for usable American pasts in the 1930s and 1940s, which includes studies about the art programs of the WPA's Federal One, the Tennessee Valley Authority, and the Section of Fine Arts, among other federal endeavors.[25] I argue that historic preservation is an important activation of usable American pasts that centers historic architecture in the narrative of New Deal cultural production.

This book also expands the political concept of New Deal federalism with regard to state building vis-à-vis public works and infrastructure construction. Scholars have examined how state and local authorities took advantage of their new relationship with the federal government to leverage public projects that improved their civic environments. In his work on the intersection of New Deal architecture and social policies, Robert D. Leighninger Jr. equates the physical building performed by federal programs such as the Civil Works Administration, the Works Progress Administration, the Public Works Administration, and the Tennessee Valley Authority with the ideological construction of liberalism and the modern economy. These programs' canals, roads, dams, airports, electrical power plants, museums, and parks critically and enduringly improved public life while strengthening the modern state.[26] Historic preservation projects, categorized within the WPA's Division of Engineering and Construction, likewise are situated within what historian Jason Scott Smith has termed the New Deal's "public works revolution."[27] Contributing to this phenomenon was the federalization of public recreation development, which Phoebe Cutler examined in her study on the parks and camps of the New Deal public landscape.[28] Attention to the historic restorations, positioned by their local and state sponsors as investments in civic and historic infrastructure, deepens this discussion of how state-sponsored public works and recreation helped materialize Roosevelt's modern liberal state.

The ramping up of federal involvement in recreational and historical tourism is also an important part of the broader context of New Deal cooperative preservation. Historians Marguerite S. Shaffer and Michael Berkowitz have explored federal interventions into national tourism during the 1930s, focusing on the National Park Service and establishment of the U.S. Travel Bureau while minorly addressing the important role local government officials, civic elites, and businesspeople played in tourism development as a strategy of economic recovery. This book reorients our focus to the nonfederal sponsors who took advantage of new federal capital and labor to develop historical tourism.[29] In his study of the Civilian Conservation Corps and Florida Park Service, historian David J. Nelson critically argues that Florida's civic elite shrewdly leveraged New Deal labor and funds to promote Florida as the nation's playground. He ultimately concludes that the goals of the local, state, and federal actors were in conflict with one another. My work provides an important interjection into this discussion, emphasizing productive ideological cooperation between federal and local partners, and the outsized role of community leaders in shaping these particular tourism ventures.[30]

Consequently, this book also contributes to the body of scholarship exploring the relationship between historic preservation and progress. Preservation historians Michael Holleran, Charles B. Hosmer Jr., James M. Lindgren, Randall Mason, and Max Page have charted the professionalization of the field, marginalization of women, and use of preservation as a tool of progressive reform in the early twentieth century.[31] Excellent recent scholarship by Whitney Martinko has expanded our understanding of the economic motivations of the preservation movement, making a compelling case that Americans in the early nineteenth century saw the cultivation of the historic built environment as a forward-thinking business venture that changed with the market. From the early national period, transactions about the future of old properties were rooted in the economic and political world.[32] These scholars provide excellent contextualization for the New Deal chapter of the preservation movement, which I call cooperative preservation, that built off of professionalized and masculinized modern management as well as the economic framing of preservation activity.

Yet preservation studies often summarily and quickly address these 1930s efforts that extended federal reach into community preservation, identifying the establishment of the National Trust for Historic Preservation in 1949 as the earliest expression of federal-local and public-private cooperation. Moreover, most studies generally consider the response to urban renewal and the passage of the National Historic Preservation Act in the 1960s as the transformative moment when the state became a critical and proactive actor in historic preservation.[33] An excellent dissertation by Lynne M. Calamia explicitly addresses how New Deal programs advanced the state's role in the preservation movement, but the work is limited to the commonwealth of Pennsylvania and focuses on the creation of a statewide historical apparatus, the Pennsylvania Historical Commission.[34] The model of federal-local cooperation was set during the New Deal era, I argue, and provided an important intermediary step before heightened federal investment in the cultural economy in the later twentieth century, which gave birth to the National Park

Service endeavors in Lowell and Boston that public historians Cathy Stanton and Seth C. Bruggeman, respectively, evaluate.[35]

Historic preservation clearly is a powerful instrument of political and economic reform as much as an experiment in historical mythmaking. As historian and architect Dolores Hayden argued, "Saving a public past for any city or town is a political as well as historical and cultural process."[36] In agreement with Hayden's assertion, I align this project with the growing scholarship on the history of public history in the United States. It joins the work of scholars who tackle the political construction of historical memory through commemorative forms like monuments, public art, museums, and other memorial landscapes.[37] Historical consciousness, a common language of symbols, and shared myths are made, debated, and remade in manipulations of place.[38] Seth C. Bruggeman reminds us that "commemoration, no matter how vigorously it nods toward consensus, is never free of argument. When memory scholars refer to the politics of commemoration, this is precisely what they are talking about: the tendency of commemoration to obscure decisions made about who gets remembered and why."[39] Scholars extensively have studied the "politics of commemoration" to expose how historic sites are shaped by the values of people with power over their preservation and interpretation. History is repeatedly collected, assembled, curated, and interpreted as expressions of political, economic, and social capital.[40] While public historians increasingly have drawn attention to the important intersection of place, historical memory, and politics, the relationship between Depression/New Deal political culture and historic preservation as a practice of public history is underexplored. Preservation scholar Daniel Bluestone explains that historic preservation, like all forms of commemoration, "involves considerable editing and selective historical recall. The process whereby individuals, institutions, and communities choose to wield both private and public power to highlight certain histories, and to ignore or render invisible others, is a critical dimension of historic preservation and public history."[41] *Restoring America* seeks to enhance our understanding of a richly textured chapter of American history by pulling back the curtain to consider which people and institutions were empowered in New Deal preservation to shape and politicize representations of the past.

Like all New Deal efforts, the preservation projects studied herein reflect efforts to revitalize American life during the Depression era. Cultural historians long have debated how to characterize the cultural and political climate of this challenging period. Ranging from Warren Susman's assessments of the era as a conservative "age of Mickey Mouse" to Michael Denning's "cultural front" of the laboring class, scholars have produced a robust and critical discussion about the extent to which New Deal culture was representative of right and left politics.[42] In this work, I follow the encouragement of cultural historian Lawrence W. Levine to "comprehend the Great Depression as a complex, ambivalent, disorderly period which gave witness to the force of cultural continuity even as it manifested signs of deep cultural change."[43] Conservative and leftist culture flourished alongside one another, and Americans often were caught in between social conservatism and political radicalism. Indeed, currents of both conservative and progressive politics reveal themselves across New Deal preservation work; the four projects addressed in this book represent escapist tendencies *and* socially critical progressive politics. For example, Connecticut's Henry Whitfield State Museum, the focus of chapter 3, aligned with nativist tenets of the popular colonial revival movement in its prioritization of the Puritan past in a tepidly Democratic state. Closer to the other end of the political spectrum is the primarily Mexican and Mexican American neighborhood of La Villita in San Antonio, Texas (discussed in chapter 5), whose restoration resulted from Democratic mayor Maury Maverick's liberal reformist agenda.

Taken together, the historic restorations discussed in this book reveal that the New Deal presents not just a narrative of controversy and contention; at times the Roosevelt administration's programs produced consensus and accommodation, between federal and local actors but also between right and left political forces. Historian Victoria Grieve uses the term "middlebrow culture" to describe New Deal enterprises designed to reach all Americans, especially the large middle class. This perspective decenters the culture produced by or for the most conservative anti–New Dealers, the poorest or most marginalized Americans, or the most radical elements of the communist left, to attend to the cultural activity that middle-class Americans performed and consumed.[44] The term "middlebrow culture" conceivably applies to the WPA and NYA historic shrine restorations and their intended audiences. They were projects conceptualized by local and state governments and civic organizations, designed by professional architects, and executed by common workers on relief registers.

Moreover, the tendency to look backward, to find relevance in history, did not necessarily express traditionalism or antimodern sentiments. The very ways in which people interacted with the historic shrine restorations required engagement in "modern" America. Visitors rediscovered these sites by reading the WPA's American Guide Series, they traveled to see them by automobile, and they witnessed relief laborers—men and women, blue-collar and white-collar alike—at work because of federal programs designed by the liberal modern state. While minor problems did surface in each location between agency officials and locals, for the most part these New Deal projects illustrated compliance and understanding, and they generally received widespread praise. This constructive relationship between federal and local actors accorded the restorations a positive political overtone and formed a productive partnership model for preservation-as-public-works endeavors. Rather than diminish their significance, the enthusiasm surrounding the restoration of historic shrines tells us that Americans critically valued the role of the built environment in the joint processes of cultural healing, economic development, and political reconstruction during the Depression.

This book is organized by case study, but before the first site takes us to Charleston, South Carolina, some orientation is necessary. Chapter 1 contextualizes the New Deal shrine restorations within the broader American preservation movement and the New Deal. It first examines the popular impulse in the nineteenth and early twentieth centuries to seek usable pasts within educational, aesthetic, and progressive reform campaigns. It then introduces the two largest, corporate-funded private preservation endeavors of the 1930s, Greenfield Village and Colonial Williamsburg, to chart how their expanded scope, methodologies, and tourism development shaped professional preservation practice. From there, the chapter provides an overview of other history-related projects initiated by New Deal programs, including the work of the WPA's arts projects, before specifically addressing the historic preservation activity of the National Park Service and work relief programs. Finally, I situate the historic shrine restorations as "construction" projects within the organizational structures of the WPA and NYA. Typical of projects sponsored by these two agencies, the four restorations were initiated by city or state governmental bodies requesting New Deal funds to either complete or begin preservation work. Throughout the projects' durations, the local sponsors and hired professionals worked in consultation with federal representatives yet enjoyed quite a lot of autonomy over the design, materials, and use of New Deal labor, illustrating the cooperative preservation that defined this era of the American movement.

The following four chapters explore the book's case studies and are organized chronologically from when the New Deal project started. Chapter 2 introduces the Dock Street Theatre in Charleston, South Carolina, a site of rich colonial lore in the South. Completed between 1935 and 1937, the WPA project embodied the desire of the city's white civic-commercial elite to recapture the romance and prosperity of "Old Charleston" during the economic turmoil of the Depression. Led by local preservation-minded architects Albert Simons and Samuel Lapham, the orchestrated "restoration" of a lost eighteenth-century theater in reality involved the rehabilitation of a once-popular nineteenth-century hostelry, which had become tenement housing by the twentieth century, and the construction of a newly designed theater inspired by 1700s contemporaries. Although a fantastical reproduction, the architects' use of salvaged elements from the Radcliffe-King House, a nearby nineteenth-century mansion, helped fabricate a visual and physical connection to Charleston's antebellum past. While the project courted conservative political interests and created a romanticized and whitewashed version of the city's history, the strong support of Democratic mayor Burnet Maybank and WPA director Harry L. Hopkins simultaneously pushed forward a progressive southern agenda to improve Charleston's economic and cultural sectors via historical tourism.

Chapter 3 moves us from a site of pre–Civil War history in the Old South to small-town New England, centering on the Henry Whitfield House in Guilford, Connecticut. Built between 1639 and 1640 as the home of the town's first Puritan minister, the so-called oldest stone house in Connecticut was a private residence for over 250 years until the Connecticut Society of the National Society of the Colonial Dames of America helped transform it into the State Historical Museum at the turn of the twentieth century. As a result of many architectural alterations, including architect-led restoration work, the house barely resembled Whitfield's home in the 1930s. A WPA project sponsored by the state of Connecticut and directed by the museum's highly involved Board of Trustees between 1935 and 1937 set out to return the house to its original appearance. Respected restoration architect J. Frederick Kelly, a regional leader in the colonial revival movement, redesigned the house to fulfill the state museum's objective: properly educate visitors about the colonial era and the town's Puritan forebearers in an accurate architectural setting. Despite complications that arose when professional standards of historical accuracy conflicted with both WPA regulations and town residents' memories of the historic property, the Whitfield House restoration persevered as an ideological commitment to the proscriptive faith and work ethic of Puritanism, and an expression of the state government's ambivalent embrace of New Deal liberalism.

The focus of chapter 4 is the early twentieth-century boyhood homestead of celebrity aviator Charles A. Lindbergh, built in 1906 on the Mississippi River in central Minnesota. Following the flier's unprecedented nonstop solo flight from New York to Paris in 1927, souvenir hunters vandalized his childhood home, a shrine of the not-so-distant past. The state first protected the simple farmhouse and surrounding lands in 1931 when it designated the Lindbergh State Park, but in 1935 the state conservation department submitted a proposal to the WPA to continue restoring the family home and further develop the park's recreational areas. The twofold project aimed to restore the pioneer landscape of an idealized Minnesota frontier, with Lindbergh and his family serving as the paragons of frontier mythology. With the WPA's involvement, however, the project moved beyond simply adulation of the famous aviator and his ancestry to advance the New Deal's conservation agenda, which favorably captured the political attention of Minnesotan farmers and the Farmer-Labor Party. Coinciding with the expansion of the modern state park system, the Lindbergh House and State Park project engaged narratives of both the state's Scandinavian heritage and agricultural history in promoting recreational use of the park and sustainable land management practices.

Chapter 5 moves us from the Works Progress Administration to the National Youth Administration and closer to the United States' entry into World War II. Between 1939 and 1942, the NYA restored the historic arts district of La Villita ("Little Village" in Spanish) in San Antonio, Texas, a once vibrant, multicultural, and socioeconomically diverse neighborhood. By the 1930s, the area had become home to mostly Mexican and Mexican American renters and deteriorated into what contemporaries called a slum. The city's Democratic mayor, former congressman Maury Maverick, used his close ties to President Roosevelt and NYA national director Aubrey Williams to make La Villita's restoration part of his sweeping reform agenda to simultaneously improve housing and sanitation, equip Mexican American youth with employable skills in arts and crafts industries, and foster relations with Latin America. Under the direction of Texan architect O'Neil Ford, NYA workers restored extant structures to the period 1722–1860, built a new museum-library building, and learned weaving, ceramics, metalwork, and more. Operating in tandem with a WPA-funded beautification project of the San Antonio River, the La Villita restoration helped boost the local tourism industry and promote a message of Pan-American identity through the preservation of the city's Spanish and Mexican architectural influences and traditional skills.

The book's conclusion briefly addresses the life of these restored shrines after the WPA and NYA laborers put down their hammers and saws before turning to some lessons we can learn from studying New Deal preservation. This work seeks to answer the question of why, during an overwhelmingly difficult chapter of American history, historic places mattered so much to so many people. What motivated these acts of historical remembrance, and what did politicians, architects, community historians, and other civic boosters hope to gain from their preservation efforts during the Depression? The conclusion addresses these questions and considers what these four projects teach us about the potential political, cultural, and economic consequences of engaging in the conservation of our historic built environment. The examples herein indisputably make clear that historic places can be public assets which play an integral role in the political life of communities as vehicles for both cultural identity construction and economic revitalization. Debates over the use of public space—about both *what* histories and *whose* histories are represented (or not), and *how* those histories are deployed for political use—are an important component of civic discourse. Examining who holds control over the physical, tangible ways history is preserved and presented demonstrates that historic preservation has been, and continues to be, incontrovertibly a political act.

On that premise, the political contours of these 1930s–1940s projects remind us that historic preservation, like all historical memory projects, reflects manifestations of privilege even if its participants aspire to create more democratic futures. I turn here, once again, to public historian Seth C. Bruggeman, who has argued that "commemoration is always political because choosing what to remember is a way of exercising power over people and ideas who are left out of our historical imagination."[45] All preservation work requires making choices to either conserve embedded power structures or change them. That American communities today face mounting challenges of economic insecurity, climate crises, and threats to political democracy makes the Depression/New Deal era a particularly apt comparative period. My hope is that within these pages there is some instruction for public history practitioners, preservationists, historians, students, and anyone who cares about historic places as to what can be accomplished when political infrastructure, sufficient and accessible funding, joint grasstops and grassroots support, commitment to inclusive and ethical history, and a shared sense of civic responsibility to manage our built heritage exist.

CHAPTER 1

America's Preservation Impulse
Before and During the New Deal

*America is a comparatively new country. Many historic buildings still
stand to tell the story of the first colorful chapters in its dramatic career
as a nation. . . . In modern designs we still use the traditional patterns
our ancestors used and we still retain many symbols of early American
culture. But in general there has been a neglect of historical landmarks
and historical records. Much material of historic significance has been lost
and more doomed to disappear as modern modes replace the old ways of
living. As an insurance against future loss and destruction, the Federal
Government, through the Works Progress Administration, is giving work
to the unemployed throughout the country on a number of projects designed
particularly for the preservation of valuable historical treasures.*

Ellen S. Woodward, director of the Division of Women's and
Professional Projects, Works Progress Administration, 1937

In 1941, American novelist John Dos Passos wrote, "In times of danger we are
driven to the written record by a pressing need to find answers to the riddles
of today. We need to know what kind of firm ground other men, belonging
to generations before us, have found to stand on." While Dos Passos referred
specifically to America's literary roots, this searching for "a sense of continuity"
with the "firm ground" of the past occurred spatially through manipulation
of the built environment during the Depression years, not just figuratively

through literature, plays, murals, and art.[1] During the boisterous 1920s leading to the Wall Street stock market crash, Americans enjoyed new forms of
entertainment and transportation like jazz, automobiles, and airplanes, which
made America's rise in economic prosperity and technological progress seem
limitless. Yet the despair and destitution of the Great Depression that characterized the following decade led people away from modern delights and
toward usable pasts in search of stability.[2]

Historian Warren Susman characterized the 1930s as a period when
Americans engaged in a "self-conscious search for a culture" and demonstrated "a commitment to forms, patterns, [and] symbols that make their life
meaningful," often looking backward in their search.[3] People explored, documented, reinterpreted, and often glorified American history to make sense
of their present moment, pursuing familiarity and seeking inspiration for a
better tomorrow in historical myths that had shaped the cultural values of
the nation.[4] Attending to the American past in both immaterial and material
forms emerged as an expression of this exploration and crafting of a national
culture during the volatile, introspective decade of the Depression. President
Franklin Roosevelt's First and Second New Deals, the sweeping set of programs designed to address the overlapping economic and social problems
and general malaise of the period, created new opportunities for people to
engage in popular history-making not previously available to them. The work
relief programs in particular offered new mechanisms vis-à-vis the political
infrastructure of federalism for local government officials, civic groups, and
professional architects and artists to activate their communities' old buildings
for economic, political, and social revitalization. Thus, the effort to ground
contemporary times in the nation's historical record materialized in increased
historic preservation activity in the 1930s.

Until those years, leaders of the historic preservation movement in the
United States were typically amateur and professional historians and civic
reformers. They were often members of hereditary, patriotic, and civic organizations who prioritized the associational histories of old buildings and
engaged in "shrine-making" commemorative efforts.[5] American preservation
blossomed in the mid-nineteenth century when middle- to upper-class white
women led fundraising campaigns, curated collections, and established historic house museums, typically with the goal of instilling in visitors patriotic
fervor and appreciation for individual buildings of the colonial and Revolutionary periods. Formal associations like the Society for the Preservation of

New England Antiquities (SPNEA) formed around the turn of the twentieth century as the colonial revival movement gained popularity, and began to both professionalize and masculinize efforts to protect historic architecture. Preservation historian James M. Lindgren detailed how this reorientation of the field introduced new modern management based around business principles, emphasized aesthetics of historic architecture over associational history, and marginalized women.[6] In the same period, the American Scenic and Historic Preservation Society, founded in New York in 1895, along with city builders advanced preservation as a *modern* strategy of progressive reform, a tool of urban improvement that was part of, not separate from, conversations about redevelopment, administrative reform, and urban beautification.[7] These late nineteenth- and early twentieth-century efforts in historic preservation generally can be characterized as localized phenomena, initiated and supported by civic groups and sometimes politicians and business communities.

The field underwent a profound transformation in the 1920s when wealthy corporate industrialists embarked on large-scale projects that catapulted historic preservation to the national stage. Two of the foremost donors and their experiments—Henry Ford's Greenfield Village in Dearborn, Michigan, and John D. Rockefeller, Jr.'s Colonial Williamsburg in Williamsburg, Virginia—greatly expanded the scope, cost, and publicity of preservation work. The two projects broadened what was considered worthy of being preserved, saving vernacular and commercial structures rather than only high-style buildings, and cemented the role of experts such as restoration architects, landscape architects, and archaeologists within the field. During the same period, the National Park Service (NPS), founded in 1916, strengthened federal commitment to administering the country's arsenal of historic places and became involved in national tourism promotion. However, the practice of preservation remained primarily a private sector endeavor with occasional support from local and state government bodies.[8]

The beginning of the Great Depression in 1929 provoked another fundamental change in the preservation movement whereby the federal government assumed a larger administrative and financial role in protecting the nation's historic resources. After Democrat Franklin D. Roosevelt assumed office in early 1933, his administration set off a series of reforms that explicitly linked work relief and historic preservation. The WPA, created by the Emergency Relief Appropriations Act of 1935, offered immediate relief aid by assigning men and women to work on projects that Roosevelt desired to be "useful—not just for a day, or a year, but useful in the sense that it affords permanent improvement in living conditions or that it creates future wealth for the nation."[9] Roosevelt's expansive definition of "useful" employment included both blue- and white-collar work performed in the agency's arts and construction projects that explored America's historical past. The WPA's shrine restorations often encompassed both kinds of labor activities. The National Youth Administration (NYA), an offshoot of the WPA, incorporated preservation and arts skills development into its youth training centers, and the Civilian Conservation Corps (CCC) often cooperated with the NPS to execute conservation work in state and national parks that called for restoring and reconstructing historic structures.

With the establishment of these New Deal work relief programs, nonfederal actors such as mayors, chambers of commerce, cultural institutions' boards of trustees, businesspeople, and professional architects discovered that federal capital, labor, and supervisory support could help make preservation a powerful form of unemployment relief and a lucrative driver of local tourism. Constituents took advantage of the new federalism of Roosevelt's modern liberal welfare state and eagerly applied for funds from work relief programs to invest in historic infrastructure. This cooperative preservation of the New Deal era advanced grassroots civic activism and created a new role for the federal government as a financial sponsor of local initiatives, establishing the basis for renewed federal participation in historic preservation in the late 1940s.

The Search for Usable Pasts at American Shrines: Patriotism, Progressivism, and Tourism

By employing the language of "shrine" to describe the restoration projects, New Dealers drew on a long tradition of framing historic places as sites of civil religion, which likened the act of architectural preservation to a secular form of worship.[10] Americans started taking "pilgrimages" to historic locations in the early nineteenth century, when commemoration efforts were usually voluntary enterprises. Churches, heritage organizations, and local historical societies raised money through membership dues and fundraising campaigns to preserve sites of important national events and the homes, especially birthplaces, of famous white men associated with the nation's founding in 1776. One celebrated example is the mid-nineteenth-century efforts of

South Carolinian Anne Pamela Cunningham and the Mount Vernon Ladies' Association to preserve George Washington's home on the banks of the Potomac River just outside of Washington, DC. When the federal government refused to protect it, the association organized to purchase, administer, and interpret Washington's estate for the public. In addition to Mount Vernon, Revolutionary-era sites like Independence Hall, Valley Forge, and Washington's headquarters at Morristown, New Jersey, became frequented historic places that glorified the struggle to forge a new nation in service of patriotism.[11] Visitation to historic shrines, argued historian John F. Sears, became a commercialized expression of secular religion that helped form a distinctly American, as opposed to European, national culture. "In a pluralistic society," historic attractions "provided points of mythic and national unity" and reflected the dominance of Protestant Christianity in the United States.[12] By the late nineteenth century, Americans regularly used "the language of spiritual transcendence to describe seemingly secular sites of public memory," writes historian Devin C. Manzullo-Thomas. An 1886 guidebook to the Civil War battlefield at Gettysburg, Pennsylvania, for example, described the revered land as "the most consecrated ground this world contains."[13] Other homes of men famous for their historical and literary deeds also achieved "shrine status" during the era, including Robert E. Lee's Stratford Hall in Virginia and Samuel Clemens's (Mark Twain's) birthplace cabin in Missouri.[14]

The Civil War of the mid-nineteenth century caused political and societal ruptures that fundamentally shifted the practices of American commemoration. In the decades following this transformative event, Americans deliberately began to craft a national historical memory shaped by their experiences and recollections of the war, as well as the processes of industrialization and urbanization. White Americans increasingly looked backward to their ancestral pasts during this period, historian Michael Kammen explained, because "the creative consequences of nostalgia helped them to legitimize new political orders, rationalize the adjustment and perpetuation of old social hierarchies, and construct acceptable new systems of thought and values."[15] People engaged in activity commemorating U.S. history, in other words, to directly address their contemporary political concerns. Rising European immigration and the resulting introduction of new foreign languages, cultural customs, and religious traditions led to the Americanization movement, an effort maintained by Protestant Anglo-Americans to "promote values that members of established families held to be traditionally American."[16] The associated colonial revival movement, in its idealization of the colonial era—quite broadly conceived as beginning in 1620 with the landing of the Pilgrims and ending around 1840 with the advent of northern manufacturing—encouraged historical activity often as a way to endorse nativism.[17]

Powerful patriotic organizations whose membership was restricted to Americans of appropriate lineal descendance, like the Society of Mayflower Descendants (formed in 1897), the Daughters of the American Revolution, and the Colonial Dames of America (both formed in 1890), devoted themselves to educating the public, especially immigrants, in the ways of Protestant Anglo-American heritage. Their methods included preserving colonial structures, erecting historical monuments, and organizing events like centennial celebrations and parades. Women's groups most powerfully exerted control over historical memory by creating historic house museums. Examples of historic sites established to advance assimilation of foreign-born Americans included the Van Cortlandt Mansion in New York City, the Betsy Ross House in Philadelphia, and the Orchard House (Concord) and House of Seven Gables (Salem) in Massachusetts, the latter of which served as an active settlement house for women and children in addition to operating as a historical museum.[18] These groups also encouraged old and new Americans from all socioeconomic stations to remember the nation's colonial history at major public attractions like world's fairs. The Centennial Exposition of 1876 in Philadelphia and the Columbian Exposition of 1893 in Chicago exhibited early American architecture, furniture, and "colonial kitchens" with volunteers dressed in period garb. Large cultural institutions responsible for interpreting national history joined more local and regional Americanization efforts in subsequent decades. The Metropolitan Museum in New York, for example, opened its American Wing of period rooms with high-style colonial artifacts and furniture in 1924, the same year Congress passed the National Origins Act restricting immigration.[19]

Leading antiquarians, architects, and both academic and community historians also formed professional societies dedicated to preserving the architectural legacy of colonial forebearers. These new groups included the Association for the Preservation of Virginia Antiquities (1889), the Society for the Preservation for New England Antiquities (1910), and the Society for the Preservation of Old Dwellings in Charleston, South Carolina (SPOD, 1920). These associations, particularly SPNEA under William Sumner Appleton's leadership, transformed the haphazardly managed field "which was ad hoc in planning, unscientific in method, and romantic in its reading of history" into a modern movement underscored by the principles of corporate organization, scientific inquiry, and male professionalism, argued James M. Lindgren. As the progressive Protestants who comprised SPNEA's membership marginalized the women long in charge of historical commemoration efforts, the preservation field "shift[ed] its orbits from romance to realism, feminine to masculine, and personal to professional."[20] However, women-led groups such as SPOD, the San Antonio Conservation Society, and garden clubs in Natchez, Mississippi, continued to wield considerable power over commemoration and preservation activities in their cities in the 1920s and 1930s.[21] Nonetheless, architectural historian Daniel Bluestone describes the general transition from women to men as part of a "redefinition of landmarks from commemoration of the past to promotion of the future, from memory to economy," emphasizing the new economic-minded management of these preservation organizations.[22]

What aided this transition were broad structural changes within the urban and industrial nation-state, including the development of a national market and transportation network, growth of a leisure economy, and expansion of national print media. These changes incentivized both private entities and the federal government to encourage national unity by developing a tourism industry centered on the nation's historic and scenic landscapes.[23] The extension of railways, building of hotels, and successful advertising of products and places created a robust vacation culture, thereby establishing a lucrative relationship between preserved historic sites and tourism. In the American Southwest and West, for example, the completion of the Atchison, Topeka, and Santa Fe route to California in the early 1880s led railroad companies to cultivate a tourist market that included hotels, restaurants, Indigenous arts and craft markets, and guided excursions through Pueblo country.[24] In New England, preservation efforts drew tourists to the recreated colonial villages of Deerfield, Massachusetts; Old Lyme, Connecticut; Cornish, New Hampshire; and Litchfield, Connecticut, where they strolled through quaint colonial town squares and admired picturesque domestic and religious architecture.[25]

In the interwar period, new developments in tourist infrastructure, especially the automobile, consciously fostered a "national tourism" that made visiting historic sites "a ritual of American citizenship," writes historian Marguerite S. Shaffer. For the first time, the federal government became "actively involved in the promotion of national tourism" when the NPS started to advertise travel to its western parks, drawing tourists away from the concentration of historical sites in New England. NPS administration negotiated with private corporations, most prominently the National Parks Association, to publicize the parks as tourist attractions and grow the burgeoning national vacation industry.[26] Vacation travel, which was once restricted to wealthy patrons with leisure time, opened up in the 1920s to salaried workers with the institution of paid time off and the increasing accessibility and affordability of the automobile. The Good Roads Association campaigned for better transcontinental highways and endorsed car travel as "an extension of America's heroic pioneer past, arguing that through the process of touring, tourists could become better Americans."[27] The rise of additional commercial enterprises associated with the automobile such as gas stations, travel planners, mapmakers, guidebooks, and auto camps meant that Americans could easily traverse the national landscape to visit historic spots both near and far from their homes.[28]

As Americans explored scenic and historic landscapes via extended national highways, private entrepreneurs and public officials actively manufactured new understandings of the historical landscape. One example is the commonwealth of Virginia's establishment of the nation's first state historic highway marker program in the 1920s, which historian Daniel Bluestone argues forged new connections "between heritage and tourism, between history and economy."[29] The federal government likewise nurtured public interest in history at the same time that it invested in road infrastructure; federal agencies including the NPS and the Bureau of Public Roads worked with state and local governments throughout the 1920s and 1930s to build the Natchez Trace Parkway linking Natchez, Mississippi, and Natchez, Tennessee; the Colonial Parkway connecting Yorktown and Williamsburg, Virginia; and the George Washington Memorial Parkway leading from Washington, DC, to Mount Vernon.[30] By the start of the Great Depression, Americans of all socioeconomic classes could partake in the modern business of tourism, whereby marketing campaigns supported by local chambers of commerce, motoring clubs, tour packagers, and the NPS beckoned people to "See America First."

In 1934, over four years into the Depression and a year into Roosevelt's First New Deal, the *New York Times* published an article by leading American museologist Laurence Vail Coleman. Coleman, who had authored the influential book *Historic House Museums* the previous year, credited the motor car with helping to save the more than six hundred historic houses across the country. "There is more than a casual connection between tourists and historic houses," Coleman wrote. "It was, in fact, the development of the automobile and the growth of motoring that gave life to the whole preservation movement." The ability to "travel casually and leave beaten paths with ease," he concluded, led to the popularity of sites far removed from train routes. He offered as examples Mount Vernon, fifteen miles outside of Washington, DC, which welcomed 500,000 visitors a year; Longfellow's Wayside Inn in South Sudbury, Massachusetts, twenty-five miles west of Boston, with 100,000 visitors; and Fort Ticonderoga in New York's Adirondack Mountains numbering over 50,000, the same as the Van Cortlandt House in New York City, located near a busy subway station. The author also noted the lure of historic sites that became resorts or retreats, early versions of B&Bs that offered guests a place to stay overnight and meals.[31] It is important to note that this growing national economy of tourism defined its market as predominantly white and middle to upper class, discouraging Black Americans and other minorities from partaking in the same enjoyment of the nation's historic sites and parks. Yet Coleman's piece clearly demonstrates that Americans' appetite for consuming history as tourists was healthy and growing when coupled with entertainment, relaxation, and validation of white national consciousness.

Corporate Efforts Shape the Preservation Field

In the 1920s, the well-funded and highly publicized historical enterprises of two wealthy businessmen, Henry Ford and John D. Rockefeller Jr., propelled the firmly established preservation movement forward and elevated it to the national stage because of their scale, expense, experimental methodologies, and tourism development. The origin stories of these two individuals' historical projects have received much scholarly consideration, but they merit some attention here for how they shaped the professional practices of the field moving into the 1930s.[32] Ford, founder of one of America's most profitable companies, the Ford Motor Company, had become a celebrity by the late 1920s for his technological innovations. Infamous for calling history "bunk," the automobile magnate embarked on an unlikely endeavor to reintroduce Americans to their past: the creation of Greenfield Village in Dearborn, Michigan, an outdoor living history museum that opened to the general public in 1933.[33]

Ford long had expressed interest in the American past and its potential role in progressive education. His first foray into preservation came in 1923 when he purchased the Wayside Inn, a site immortalized in Henry Wadsworth Longfellow's 1863 collection of poems *Tales of a Wayside Inn*. As a precursor to Greenfield Village, Ford gathered colonial buildings from across New England and restored them to their eighteenth-century appearance to create Old Sudbury, a 2,500-acre colonial village in eastern Massachusetts with the Wayside Inn as its centerpiece.[34] Ford, himself the son of an Irish immigrant and both xenophobic and antisemitic, established the Wayside Inn School for Boys to teach foreign children in compulsory citizenship and language classes. The school, according to historian Jessie Swigger, "exemplified Ford's signature mix of progressive and traditional pedagogies by offering hands-on instruction in a colonial setting."[35] This venture to connect preservation with Americanizing efforts placed Ford within the popular colonial revival movement, which he actively advanced during his twenty-three years as SPNEA's vice president. It is fitting, then, that he selected Independence Hall as the model for the main facade of the Ford Museum, also in Dearborn, Michigan, which opened five years before Greenfield Village in 1928.[36]

At Greenfield Village, Ford curated an artificial historic American community. The 260-acre village represented a hodgepodge of structures from the seventeenth to twentieth centuries, visually narrating a history of American technological progress and scientific discovery that privileged narratives of individualism, entrepreneurship, and business. By the time of his death in 1947, he had relocated and/or replicated eighty-six buildings, which included the Noah Webster House, the Wright Brothers' Cycle Shop, and Thomas Edison's Menlo Park Laboratory. Altogether, Greenfield Village reflected Ford's idealization of small-town life, his faith in the hard-earned upward mobility of entrepreneur-inventors, and his predilection for technological and industrial history.[37]

Historian Karal Ann Marling vividly described Ford's pet ensemble of historical structures as "wildly and gloriously wrong, a potpourri of gripping moments, patently quaint sights wrenched from any page of history ripe for the pillaging, without much regard for temporal distinctions between colonial times and the life and times of Henry Ford."[38] Yet Marling's criticism underscored Ford's motivation for his odd accessioning of buildings: he intended Greenfield Village to be an "animated textbook" for the broad American public, a teaching laboratory where people could see, touch, and hear history that felt relevant and palpable. History-learning became experiential in Dearborn, not rote memorization.[39] While Ford's uprooted and assembled village raises questions about historical authenticity and integrity, his methodology left a mark on the field. Indeed, historian Steven Conn argued that "what is perhaps most significant here is not the particulars of Ford's historical fictions, but the way he tried to tell the story—not the product but the method."[40]

Importantly, Ford adamantly called for the expansion of the study of history—in his view a stuffy, dry, scholarly field—to include everyday artifacts that spoke to common American experiences. To Ford, things were more powerful than words in interpreting the past because they were tangible representations that communicated directly with the public. While historical societies and museum professionals already were arguing that material culture of ordinary life was important to historical understanding, Ford's inventive project helped to popularize this interpretation. Yet, crucially, Ford's laudatory view of common objects and vernacular architecture did not encompass the importance of *place* in the contextualization of where events happened or people lived and worked. Ford believed historic buildings articulated the power of their past even if divorced from their original setting.[41] Despite the limitations of this object-based epistemological perspective, he undeniably made a powerful argument that buildings could teach as effectively, if not more so, as books, helping spark the living history movement that grew dramatically in the next couple of decades with the establishment of outdoor museums such as Old Sturbridge Village and Historic Deerfield in Massachusetts and Old Salem in North Carolina.[42]

The second large-scale private preservation endeavor of the time that drew national acclaim was the restoration of Williamsburg, the capital of colonial Virginia. During the early twentieth century, members of the Association for the Preservation of Virginia Antiquities functioned as the "keepers of the past" in Williamsburg, and the Tidewater town remained a quiet, fairly neglected place.[43] But the efforts of an Episcopalian priest and America's wealthiest industrialist-cum-philanthropist altered both the stewardship and prestige of the historic city. In the mid-1920s, Reverend William A. R. Goodwin successfully implored John D. Rockefeller Jr. to restore Williamsburg to its colonial appearance, thereby transforming the area into a premier outdoor history museum. Opening in 1934, just a year after Ford's Greenfield Village, Colonial Williamsburg soon became a leading example of technical preservation practice. It adopted many of the practices already institutionalized by

SPNEA in its local and regional work—scientific research, expert guidance, and business-minded management—but the professionalism, depth of architectural study and archaeological excavations, and holistic approach to restoring the town of Williamsburg as a cultural landscape rather than focusing on select individual buildings set national standards for future private and federal preservation work.

In initially promoting the Williamsburg restoration, Goodwin employed the by-then common language of civil religion to position the project as a shrine to Americanism. He told Rockefeller that at Williamsburg they could "reproduce the symbols and sacraments of the past."[44] Like members of patriotic organizations who established house museums in the preceding decades, Goodwin and Rockefeller believed Colonial Williamsburg could reacquaint Americans with the tenets and values of the nation's Anglo founders, a spiritual panacea to the perceived dangers of foreign influences like socialism and anarchy. Historian Anders Greenspan contends that the Williamsburg restoration gave visitors "a chance to physically interact with a sanitized version of the past and to gain a better appreciation" for Rockefeller's and Goodwin's specific notion "of what it meant to be an American," which was as narrowly defined along white Protestant lines as was Ford's idea of a true American.[45] Visitor reports and the press revealed that the town indeed had been popularized and propagandized successfully as a pilgrimage site. The *Virginia Gazette*, for instance, called the restored village a "mecca for the thousands who wish to see this sacred and historic spot."[46]

While both Ford and Rockefeller believed in the educative qualities of their respective projects, the Williamsburg restoration differed in its intentional focus on historical accuracy and authenticity. This scientific approach marked an important departure from Ford's curated and idealized colonial village in Dearborn but aligned with the professional work of architects hired by organizations like SPNEA and the Colonial Dames of America. Kenneth Chorley, president and director of Williamsburg Restoration, Inc., summarized the team's preservation tactic in 1941: "Authenticity has been virtually the religion of our institution, and sacrifices have been offered before its altar. Personal preferences, architectural design, time, expense, and, at times, even the demands of beauty have given way to the exacting requirements of authenticity."[47] Rockefeller and the professionals he employed, which included architects, archaeologists, and photographers, wanted visitors to believe they were seeing an authentic representation of the eighteenth-century colonial capital. To produce this effect, workers conducted extensive archaeological excavations; reconstructed prominent colonial sites that had been demolished, like the Governor's Palace, Capitol Building, and Raleigh Tavern; and restored the extant Courthouse, Public Magazine, and George Wythe House. These efforts to recreate the eighteenth-century town also called for the demolition of no fewer than 790 structures and the removal of any traces of post-1800 architecture from the remaining 82 houses.[48] To protect the professional integrity of the restoration, Goodwin proposed the creation of a board of experts in the fields of "Architecture, creative Literature, History, Education and Interpretation."[49] This move toward objective, professional, and outsider evaluation indicated a significant contrast to Greenfield Village, where Ford's thumbprints were evident all over the decisions about the site's administration and interpretation. In prioritizing expert opinion, Williamsburg served as a model for the NPS when it developed new agency restoration policies in the 1930s.

Also noteworthy is the Rockefeller-funded restoration's role in providing jobs to local workers during the Depression, a time when unemployment across the country skyrocketed. While some of the steel workers hired for the job came to Williamsburg from New York, the majority of the carpenters, brick masons, and plasterers employed on the project were local residents. In 1937, the site established craft shops that emphasized the art of historic trades and the importance of the individual artisan, simultaneously endorsing Goodwin's and Rockefeller's credence of American economic individualism. The Colonial Williamsburg craft shops mirrored the craft training centers established by the NYA during the same period, although the New Deal sites focused specifically on equipping young workers between the ages of sixteen and twenty-five with employable skills.[50] In its early years, Colonial Williamsburg received praise from news outlets, visitors, and the federal government alike for its role in providing economic opportunities to the local population, bringing national attention and tourists to the small Tidewater village, and prioritizing eighteenth-century white history at a time when the prevalent colonial revival movement captured political conservatives' aesthetic attention.[51] Rockefeller's labor of love had turned out to be a remarkably successful business endeavor, historical project, and experiment in preservation as a form of community revitalization.

While advancing the preservation movement, both Greenfield Village and Colonial Williamsburg reinforced a white middle- to upper-class hegemonic narrative about the American past. Their recreated environments transported visitors to celebrated moments in white U.S. history, and reproductions of shops, homes, and colonial activities instructed Americans on the way things used to be according to a particular historical viewpoint. Contributing to both projects' popularity were their glaring absences. For example, the extensive efforts to restore the eighteenth-century Virginian capital led to the displacement of large numbers of people living in the town, including many Black and working-class residents.[52] At Greenfield Village, Ford prioritized vernacular architecture to express the historical realities of ordinary Americans, yet he ignored contemporary conflicts between labor and capital, disassociating the industrial past from his own business practices.

Nonetheless, the scale, publicity, and popularity of these two preservation projects in the 1930s further joined pilgrimages to historic shrines with the burgeoning tourist industry, and set the groundwork for New Deal preservation projects that positioned restoration as a tool of local economic recovery. They created educational and entertaining historical landscapes that could be traversed by automobile, feeding Americans' appetite to spatially consume historical content and spend tourist dollars. Various New Deal programs built off of this desire to seek accessible pasts and employ local workers in both their cultural and construction projects, linking preservation with the federal government's economic relief programs.

The New Deal Cultural Programs Explore the American Past

The stock market crash of 1929 launched Americans into a full-scale economic and cultural crisis that President Roosevelt, after his inauguration in 1933, endeavored to manage through his sweeping New Deal reform program. His administration's cultural agenda tapped into the by-then firmly established trend to use historic narratives and sites as ameliorative forces. The New Deal cultural programs most popular for both their production and political notoriety were the five units that comprised the WPA's Federal Project Number One (Federal One): Federal Art Project (FAP), Federal Theatre Project (FTP), Federal Music Project (FMP), Federal Writers' Project (FWP), and the Historical Records Survey. From 1935 to 1943, with the exception of the FTP, which Congress shut down in 1939, Federal One produced a tremendous outpouring of American art in various forms, including plays, folk songs, travel guides, films, novels, slave narratives, and more. While the primary goal of Federal One was to employ out-of-work artists, the ideological imperatives of the programs' administrators, themselves leading experts in artistic and literary fields, engendered cultural production that dismantled the hierarchical superiority of high culture and brought art to ordinary Americans.[53] Cultural workers who traditionally had to rely on private commissions from wealthy patrons to create and present their work now had opportunities to produce art for public buildings and broader and more diverse audiences through federal employment. National administrators such as curator and writer Holger Cahill (FAP), musicologist Charles Seeger (FMP), and journalist Henry Alsberg (FWP) embraced the search for usable pasts by assigning workers the tasks of studying and documenting the repertoire of American folklore and art.[54]

The theme of exploring U.S. history to make sense of the present was especially prevalent in the more than 1,100 post office murals commissioned by the Treasury Department's Section of Painting and Sculpture (later renamed the Section of Fine Arts and then moved to the Public Buildings Administration of the Federal Works Agency) between 1934 and 1943.[55] While not sponsored by a Federal One program, the section's so-called American Scene murals similarly prioritized "the people," emphasized local and regional environments, and highlighted uplifting local historical events and narratives of progress within the art movement of American regionalism.[56] American Scene paintings often hearkened to European American pioneer history and foregrounded inspirational figures like frontiersman Daniel Boone, explorers Meriwether Lewis and William Clark, and log-cabin-born president Abraham Lincoln, much like popular historical writing of the period. Collectively, the murals projected messages of correlation and continuity between past and present, suggesting that there were abiding values and stability to be found in the examination of regional and national heritage.[57]

An especially popular New Deal initiative that prioritized local history and moved the built environment to a starring role in celebrations of American culture was the FWP's American Guide Series. Between 1937 and 1941, FWP writers produced over four hundred guidebooks, including volumes for each of the forty-eight states plus the territories of Alaska and Puerto Rico; New York City (two guides); the District of Columbia; four major U.S. highway routes, including the Ocean Highway and Route 1; and a number of local guides.[58] The state guidebooks, the initiative's most widely read publications, were divided generally into three sections: the first third presented essays on local history, geography, architecture, economics, and culture; the second offered itemized descriptions of towns and cities; and the third featured detailed automobile tours of prominent historic and nature destinations.[59] In his analysis of the cultural work of the FWP, historian Jerrold Hirsch contends that the guidebook essays on art and architecture, supervised by national FWP editor Roderick Seidenberg, "were markedly superior in style and content" to other essays on the arts.[60] Seidenberg viewed the guides as instructive literature that reacquainted Americans with the built environment by illustrating "a broader conception of architecture as an expression of historic (and social) forces—as a resolution, in visible form, of the trends and tendencies of our civilization."[61]

FWP national director Henry G. Alsberg hoped this reenergized appreciation for history written into spatial surroundings would help "prevent relics of the past from crumbling into dust" by cultivating "a sense of local pride of possession."[62] The series taught citizens to view historic architecture—the vernacular and the high-style alike—as local shrines for which they had proprietary rights and responsibilities, a mindset already popularized through the historic house museum movement. Although in written form, the guides attached compelling stories to buildings, linking idiosyncratic and uplifting historical narratives to extant structures. As Hirsch explains, "Rather than erecting memorials to locate the past in the present, the FWP writers tried to make the existing American landscape into a monument, to infuse it with emotion and symbolic content."[63] By using the guidebooks, Americans rediscovered the diverse, storied, and manipulated landscapes that constituted their historical geography, whether through consumption by reading or physical trips in their automobiles.

Despite their apparent uniformity and widespread positive reception, the guidebooks are more complicated cultural products than they appear at face value. Scholar Christine Bold argued that the guidebooks were not "unproblematical celebrations of American democracy and cultural diversity. . . . [They] expose the complex cultural processes set in train by federal intervention into local image-making, the cultural fallout from the New Deal mapping of public space."[64] The FWP writers' assessments of architecture, though positing the built world as an expression of cultural forces, did not encourage readers or travelers to ask "questions concerning process and pattern," leading to one-dimensional narratives of "indigenous mythic history."[65]

As products of an ideologically driven federal project, the American Guide Series instructed tourists on how to read the landscape by telling them where to go, what to see, and in what order. Thus, the guides served as a form of government propaganda meant to instill a sense of pride and patriotism in their eager audiences.

To that point, American writer, urban planner, and architectural critic Lewis Mumford praised the guidebooks in 1937, calling the series "the finest contribution to American patriotism that has been made in our generation. . . . Let it also silence those who talk with vindictive hooded nods about the subversive elements that are supposed to lurk in the WPA."[66] In fact, sociologist Wendy Griswold shows how the guides were created to stave off the three main criticisms of Federal One, chiefly that the WPA cultural projects were "boondoggles," or projects that wastefully spent federal money; that Big Government inserted itself excessively in local and state affairs; and that the arts projects harbored communists and radicals who produced anti-American propagandic art. Publicity about the American Guide Series minimized the political underpinnings of the project, disconnecting them ideologically from the New Deal work relief program out of which they sprang.[67] By focusing on tourism and exploration of the American landscape, the guides found mass appeal, even among the WPA's strongest critics within the business community who otherwise attacked Roosevelt's programs.[68]

The FWP's American Guide Series was not the only expression of federal intervention into history-based tourism during the Depression. In 1937, the government initiated what historian Michael Berkowitz calls its "most significant contribution to the field of tourism promotion and coordination."[69] U.S. Secretary of the Interior Harold L. Ickes created the U.S. Tourist Bureau (USTB; later the U.S. Travel Bureau) within the National Park Service using funds from the WPA and CCC. President Roosevelt later signed legislation creating the USTB as an independent office within the Department of the Interior with its own appropriation in 1940. Its purpose was to encourage domestic tourism to spur economic recovery, ignite patriotism, strengthen belief in American democracy amid looming war in Europe, and endorse environmental conservation. Like the American Guides but to an even greater extent, the USTB's utilization of tourism as economic development appealed to even the most resolute critics of Roosevelt's New Deal. The private sector not only supported the USTB but was instrumental as lobbyists in Congress to pass the bill creating it and later to increase its funding.[70]

The USTB "assumed the role of a national clearinghouse" and collaborated with touring companies to disseminate tourist information from its offices in New York, Washington, and San Francisco. Through newsletters, radio programs, and special travel bureau posters published by the WPA, it encouraged citizens and foreigners alike to spend tourist dollars. The USTB advertised the New York and San Francisco World's Fairs of 1939 and campaigned on behalf of Roosevelt's "Travel America Year" of 1940. Although the USTB was a short-lived agency with World War II leading to its closure in 1943 (before it restarted in 1948 as the U.S. Travel Division), Berkowitz argues that the combined efforts of the U.S. government and community-advertising organizations to sell tourism to residents of their home and distant communities "fundamentally shaped the cultural life of New Deal America while simultaneously helping to create a commercialized, profit-driven industry of enormous economic power."[71]

The Roosevelt Administration Expands Federal Preservation Work

The creation of the Civil Works Administration (CWA) in November 1933 presented the first opportunity for the federal government to promote and finance preservation activity through a work relief program. Designed to generate jobs for out-of-work Americans, the CWA encouraged federal agencies to submit proposals for programs that could alleviate the unemployment problem. One proposal came from Charles E. Peterson, chief of the NPS's Eastern Division of the Branch of Plans and Designs. Earlier that summer, Roosevelt's Executive Orders 6166 and 6228 transferred all sixty federally controlled parks, monuments, cemeteries, and battlefields from the War and Agriculture Departments into the Department of the Interior's NPS. With the stroke of a pen, this quadrupled the number of historic areas under NPS supervision.

Peterson, responding to the agency's larger role in preserving the nation's historic sites, envisioned a six-month work relief program employing one thousand architects, photographers, and draftspeople to complete a national survey of American architecture through measured drawings and photographs. NPS officials quickly approved the proposed program, called the Historic American Buildings Survey (HABS), which also found fast support from Secretary of the Interior Harold Ickes.[72] When the CWA ended in

July 1934, the NPS, the American Institute of Architects, and the Library of Congress signed a memorandum of agreement to ensure HABS's continuation. Under this tripartite agreement, architects under the auspices of the American Institute of Architects identified and catalogued the structures, NPS employees photographed buildings and prepared measured drawings, and the Library of Congress served as the repository for the inventory forms, drawings, and photographs.[73] With its formal and permanent establishment, HABS later became the WPA's Federal Project Number Two and provided the historic preservation movement with what the private sector could not, according to architectural historian Annie Robinson: "the constituency, the publicity, and the money to create a national archive of measured drawings of American architecture."[74]

Rather than catalogue buildings only of singular value or high-style designs, Peterson argued for recording "structures which would not engage the especial interest of an architectural connoisseur," including "the great number of plain structures which by fate or accident are identified with historic events."[75] Under this mandate, HABS became a remarkably inclusive program, documenting not only elite or exceptional buildings but endangered vernacular structures like bridges, barns, forts, and mills, focusing especially on those built before 1860. Ellen S. Woodward, director of the WPA's Division of Women's and Professional Projects, favorably described the diversity of documentation work of HABS draftspeople in 1937:

> With cameras, tapes, compasses and drawing boards, the workers are gathering all possible data on such dwellings as the adobe hut, the Indian tepee and cliff houses, the pioneer log cabin, the cottage, the farmhouse and the city residence. Over the entire United States they are recording buildings which possess exceptional historic interest, particularly those that are in danger of being destroyed. Into the architectural record go the churches and missions of the Franciscans and Jesuits of the South and West, the churches of the Russians in Alaska, the meeting houses of the Puritans in the East and Middle West, the college, hospitals, mills, covered bridges, shops and other structures of social as well as architectural significance.[76]

Woodward expressed similar admiration for the historical work of the FAP's Index of American Design, a compilation of more than eighteen thousand watercolor renderings of American folk, decorative, and commercial art and objects from the colonial period through 1900, and the WPA's Historic American Merchant Marine Survey, a project sponsored by the Smithsonian Institution that hired marine architects and surveyors to record significant vessels.

Overall, in the eight years before the United States entered World War II in 1941, HABS employed approximately 1,600 architects and draftspeople who produced an impressive archive of 23,765 drawings and 25,357 photographs of more than 6,000 buildings in 42 states, the District of Columbia, and Puerto Rico.[77] Annie Robinson has argued that HABS's "preservation-through-drawing methodology complemented the bricks-and-mortar approach to rescuing buildings" of private sector preservation projects.[78] Moreover, as historian Michael Wallace observed, HABS workers documented buildings whose "historical importance was rooted in local memories and traditions."[79] The program's emphasis on structures built by common builders and made using local and Indigenous methods aligned with the democratizing agenda of the aforementioned Federal One arts program, as well as the broader documentary impulse of the 1930s.

The NPS cooperated with other New Deal programs besides HABS to initiate or complete archaeological, conservation, and construction projects, and provided technical and administrative assistance for immediate and long-range park developments in both national and state parks. Before the New Deal programs ended, the NPS worked with at least five agencies, including the Federal Emergency Relief Administration (FERA), CWA, Public Works Administration, WPA, and CCC, which put single young men to work improving the nation's public lands, forests, and parks.[80] During its nine years of operation from 1933 to 1942, CCC enrollees, under NPS supervision, completed reforestation work, insect and disease control, and erosion management, and they built fire trails, lookout towers, ranger cabins, sanitation and water systems, housing for employees, service roads, and recreational structures like park shelters. The young CCC workforce also regularly restored historic buildings, designed museums and exhibits, and completed minor archaeological projects at parks.[81] As historian Patricia West has noted, the CCC "hired young professionals as 'cultural foremen' or 'history assistants' to work for the budding NPS history program."[82] The CCC was particularly active in the restoration of Civil War sites, including Shiloh, Petersburg, Chattanooga, Chickamauga, Spotsylvania, Fredericksburg, and Appomattox.[83] Additionally, the NPS, U.S. Forest Service, and the CCC–Indian Division co-developed a preservation project to employ Alaskan

Native craftspeople to conserve Indigenous cultural assets. Enrollees restored or reproduced hundreds of totems between 1938 and 1942, including totems at the Chief Shakes Historic Site, Sitka National Historical Park, and Klawock Totem Park.[84] Other New Deal programs supported historical archaeological work across the United States, particularly the CWA, WPA, and Tennessee Valley Authority.[85]

During the first year of the Second New Deal, Congress significantly increased the NPS's role in the protection of both historical and archaeological sites with the passage of the Historic Sites Act of 1935. It became the nation's first comprehensive historic preservation bill and broadened the scope and responsibilities of the preexisting Antiquities Act of 1906.[86] Inspired by the success and popularity of Colonial Williamsburg, former NPS director Horace M. Albright persuaded his friend John D. Rockefeller Jr. to fund a comprehensive study of preservation legislation and activity in both the United States and Europe.[87] The result of this study was the influential *Report to the Secretary of the Interior on the Preservation of Historic Sites and Buildings*, which Ickes used to draft a landmark national historic preservation bill. According to Patricia West, the report "pointed to millions of yearly visitors to hundreds of house museums and bemoaned the small role that the national government was playing in the phenomenon."[88] While the government had become more active in national tourism ventures, its direct participation in preservation activity remained minimal and mostly confined within the NPS.

Increasingly politicians, the business community, and civic organizations called for more federal oversight in historic property management. Texan congressman Maury Maverick, who we will meet again in chapter 5, introduced the bill to create the Historic Sites Act in the House of Representatives in March 1935, two weeks after it was introduced in the Senate by Virginia's Harry F. Byrd. Ickes enthusiastically promoted the bill, recognizing the swelling appreciation for historic places amongst Americans: "In the past few years the American people have displayed a sharply increased awareness of its historic past. This growing interest and pride in both local and national history is a healthy and encouraging phenomenon which is reflected in the every-increasing number of bills being introduced in both Houses of Congress, providing for the marketing, preservation or restoration of historic sites or structures throughout the country."[89]

Less than one week after the congressional hearings approved the legislation, President Roosevelt indicated his full support for the proposed act: "The preservation of historic sites for the public benefit, together with their proper interpretation, tends to enhance the respect and love of the citizen for the institutions of his country, as well as strengthen his resolution to defend unselfishly the hallowed traditions and high ideals of America. At the present time when so many priceless historical buildings, sites and remains are in grave danger of destruction through the natural progress of modern industrial conditions, the necessity for this legislation becomes apparent."[90] Tending to the nation's historic places, Roosevelt argued, was a powerful endorsement of American patriotism and a commitment to democratic institutions, a philosophy that aligned with his administration's broader New Deal agenda as well as the large-scale philanthropic preservation work of the 1920s and 1930s.

When Congress officially approved the Historic Sites Act in the summer of 1935, the NPS truly became a proactive federal vanguard of the historic preservation movement. This act provided for "the preservation of historic American sites, buildings, objects, and antiquities of national significance," and declared it "national policy to preserve for public use" historic places "for the inspiration and benefit of the people of the United States." The law outlined the magnified powers and duties of the secretary of the interior, enacted through the NPS, which included the authority to conduct historical research and documentation, perform survey work, acquire historic property in the name of the United States, enter cooperative agreements with government entities or individuals, erect historical commemorative markers, operate historic sites and perform preservation work, and develop an educational program to disseminate information to the public. The act also created the Advisory Board on National Parks, Historic Sites, Buildings, and Monuments, composed of eleven experts appointed by the secretary in the fields of history, archaeology, architecture, and human geography.[91]

From 1935 to 1937, the NPS's Branch of Historic Sites and Buildings consulted with the Advisory Board and technicians from other branches to develop a "proper restoration policy" for new units added to the system. In March 1937, the Advisory Board approved a memorandum announcing new NPS policies for general restoration and battlefield area restoration, along with a sample policy of the Morristown National Historical Park in New Jersey, which took effect on May 19, 1937 (see appendix B).[92] The "General Restoration Policy" acknowledged that the "aesthetic, archaeological and scientific, and education" motives guiding preservation often conflicted. To steer the "tact and judgment of the men in charge" while "attempting to reconcile these claims and motives," the policy offered nine general guidelines. They included exhaustive research of archaeological and historical evidence before taking any course of action; complete recordkeeping of said evidence; the study of comparative examples from similar periods and regions when replacing evidence; and expenditure of reasonable care, cost, and time to produce quality work. The third guideline coached NPS staff, "It is well to bear in mind the saying: 'Better preserve than repair, better repair than restore, better restore than reconstruct,'" while the sixth guideline cautioned, "In no case should our own artistic preferences of prejudices lead us to modify, on aesthetic grounds, work of a bygone period representing other artistic tastes."[93] These new policies emphasized technical work, architectural integrity, and objective evaluation, bringing the NPS's preservation methodology into accord with the professionalized and scientific trends established within the private sector.[94]

Historic Shrine Restorations as New Deal Construction Projects: (Re)Building the Modern State through Cooperative Preservation

By the mid-1930s, it was clear that historic preservation had moved solidly from a private sector issue to a public sector concern with the expansion of the National Park Service, passage of the Historic Sites Act, and implementation of various New Deal programs' historical, conservation, and archaeological projects.[95] Indeed, an article published in the *New York Times* in 1939 described the national preservation impulse as a "cultural renaissance of the country's appreciation of its heritage" that was "given government impetus" in 1935 through two particular events, citing the approval of the Historic Sites Act and the establishment of the Works Progress Administration.[96] The WPA and NYA restoration projects that are the subject of this book are a consequence of this "renaissance," a result of the heightened public interest in history and usable pasts, expanded federal involvement and policies in the preservation field, and availability of New Deal funds that enabled elected local leaders and civic groups to initiate historic landmark restoration as public works projects.

The bureaucratic position of the historic shrine restorations within the WPA's organization defined the interplay between federal and local actors and their political reception at all governmental levels. The restorations fell under the agency's Division of Engineering and Construction, which was classified into six categories: municipal engineering projects, public buildings, airports, highway and road projects, conservation projects, and engineering surveys. Preservation projects typically belonged in the first category—there was no separate branch for restorations—and thus joined the relief program's impressive and diverse inventory of auditoriums, roads, bridges, parks, libraries, schools, and other public facilities. WPA projects had to be on public property—the agency did not allow for work relief construction on private land—and they had to have a federal or nonfederal government sponsor.

The vast majority of WPA projects had nonfederal sponsors, meaning that state or local governments more familiar with local conditions and better positioned to determine community needs proposed the projects and submitted applications to the agency's central office. Private organizations like civic clubs, churches, or veterans' associations could not directly sponsor projects, but these groups often brought preservation concerns to the attention of government offices and coordinated with them throughout.[97] Once the projects were completed, it was the sponsor's responsibility to maintain and operate the sites at their own expense, thereby shifting the financial and administrative management from the federal to local or state actors. The restorations of both the WPA and the NYA, then, were transient projects of a work relief program with defined timeframes rather than permanent historic sites under the perpetual responsibility of the federal government, like units of the NPS.

The application procedure for WPA project approval was uncomplicated. The sponsoring agency completed a basic form (form 301) that included a project description; detailed budget with cost estimates for labor, supervision, and materials expressed in contributions from both the federal agency and the sponsor; a breakdown of labor type into unskilled, intermediate, skilled, professional, technical, and supervisory; a list of required equipment and materials; the estimated monthly employment numbers; and a justification statement for the proposed project. For all construction projects, engineering plans and sketches had to be submitted along with the applications. The application first went through the state WPA office's Division of Engineering and Construction, which either approved the projects or returned them for revision, and then the state administrator sent a formal application to the central DC office. Once the central office authorized the project, which included getting approval from an Advisory Committee on Allotments and President Roosevelt, the WPA district director coordinated with the sponsoring agency to allocate personnel from relief registers and select the forepeople and time-keepers. About 90 percent of workers were to be taken from local relief registers, and district and state offices determined relief eligibility.[98]

During the WPA's eight-year existence from 1935 to 1943 (including after Congress reorganized the agency and changed its name to the Work Projects Administration within the newly established Federal Works Agency in April 1939), the federal government typically provided 75–80 percent of the cost of projects, which generally funded the labor, while the sponsor contributed the remaining 20–25 percent, covering the costs of materials, tools, and any specialized skilled labor. This standard funding distribution varied depending on the project and the amount of resources available at a given time, which sometimes resulted in the WPA assuming a larger percentage of the total cost.[99]

On June 26, 1935, President Roosevelt announced the creation of the NYA within the WPA. The government later transferred the NYA into the Federal Security Agency in 1939 following passage of the Reorganization Act, and moved it again to the War Manpower Commission in 1942. The NYA program provided work and education to young Americans between the ages of sixteen and twenty-five in four categories of projects: service and clerical work, recreational improvements, public works programs, and resident training centers.[100] Like its parent agency, the NYA operated as a decentralized national program, with forty-eight state offices supervised by a central office in Washington. It followed a similar application process and also usually required that 75 percent of all funds spent on a project went toward wages, while the remaining 25 percent were allocated for materials, equipment, and supervision.[101] In Texas, the first state NYA administrator was future president Lyndon B. Johnson (1935–37), followed by Jesse Kellam (1937–43), who worked with Mayor Maury Maverick to restore San Antonio's La Villita.[102] When Johnson held the position of Texas state administrator, he established over two hundred state, county, and local advisory committees to provide guidance to the state directors, recognizing that appointed citizens would know best how to support the needs of youth in their respective areas. These committees advised on the direction of projects and helped locate sponsors and funding.[103]

Administratively and functionally, the restorations' categorization as construction projects firmly positioned them as agents of economic recovery. When considering the intellectual labor they required, however, the preservation projects aligned conceptually with the white-collar work of the WPA's

Federal One (FAP, FMP, FTP, FWP) and Two (HABS). The American buildings and cultural landscapes preserved as shrines added to the growing portfolio of national art that emphasized regionalist and folk traditions rooted in local history and culture. Yet, the restorations' institutional categorization as construction projects meant they differed from the WPA's cultural programs in important ways that shaped their origins and the political dynamic between local and federal sponsors. Unlike the rest of the WPA, the Federal One projects' state directors did not report to the WPA state administrators but directly to national administrators in DC. However, the Federal One programs still depended on state administrators for assistance, which led to significant tension between DC staff and state offices. In her study of the FWP's American Guides, Wendy Griswold explains how strained relationships among national, state, and local administrators led to debates over FWP publications' formats, content, language, and authors, which resulted in their "peculiar blend of conventional and idiosyncratic. . . . Each [guide] bore traces of the struggle between Washington standardization and state individualism."[104]

In addition, the cultural work the arts programs produced was highly politicized—often more so than projects in the Division of Engineering and Construction—because of their suspected associations with leftist and radical culture. The perception of the FTP's plays as politically threatening, for example, led to the program's early end through the Emergency Relief Act of 1939, which also moved the three other arts programs under state rather than federal oversight, ending the disproportionate control arts directors wielded over projects in the states.

The WPA's Division of Engineering and Construction, in contrast to the arts programs, operated with a fairly decentralized and procedural structure organized at four administrative levels—the central office in Washington, DC; the regional offices; the state administrations; and the district offices. The lowest level of the bureaucracy, the district offices, was responsible most directly for project operations. For example, WPA district officers assigned personnel from local relief registers, named project superintendents, set timelines, published progress reports, checked for engineering soundness and legality, and cooperated directly with local sponsors.[105] Because of this organizational structure, local politics and cultural ambitions largely guided the restorations rather than federal administrative mandates, political objectives, or individual administrators' undue influence. While WPA staff at the district and state offices held roles varying in degrees of importance and involvement from project to project, the sponsors of the historic shrine restorations and other invested figures—be they local or state politicians, boards of trustees, state conservation departments, professional architects, civic and business elites, or other townsfolk with a deep interest in local history—were highly influential in determining the political, economic, and aesthetic contours of the projects. As will be seen in the following chapters, state WPA and NYA administrators generally played a minimal role, serving more as advisors than active participants in the design and direction of the shrine projects. There was, in fact, remarkably little tension between New Deal administrators and local sponsors in the projects covered herein, which consequently kept the restorations in good political standing.

Local and state governments, urged by politicians, civic groups, and cultural institutions, submitted applications to the WPA to fund preservation projects in order to leverage federal support for community economic revitalization. They viewed the work relief program as a way to wrest control of money, labor, and property to enact municipal agendas, and by doing so marked a shift in the field toward federal-local cooperation. Before the mid-1930s, preservation activities, supported and funded by the private sector, steered clear of federal involvement and focused primarily on the educational, aesthetic, and tourism possibilities of historic sites. Corporate preservation like that of the Ford and Rockefeller endeavors, while helping to professionalize and popularize preservation, were guided by benefactors' personal and narrow ideologies about American history and the nation-state. The New Deal enabled local government officials and civically engaged residents to articulate and execute their own visions of preservation as a tool of reconstruction, giving a voice to community interests in preserving historic places. Moreover, because of the precedent of active federal involvement in tourist promotion, including the efforts of the NPS and the establishment of the U.S. Travel Bureau, the historic site restorations faced far less criticism than other WPA projects, and, as is illustrated in the next chapters, often had nonpartisan support and involvement from the business community.

In 1938, writer and English professor Charles I. Glicksberg lauded the impressive construction projects of New Deal programs as "useful contributions to the material welfare of the nation." But it was the federal arts projects, he contended, that brought "cultural renewal" to the country as "the things of the spirit generally have a more lasting memorial." He continued, "There can[not] be continuity of tradition unless a nation takes pride in preserving the records and memorials of the past. American communities have been culpably careless in this respect. Whatever work of preservation has been done is largely due to the efforts of historical societies and university libraries in various cities. . . . The healthy growth of a national culture depends in part on this sense of kinship with what has gone before, an awareness of the rich foreground."[106] The WPA and NYA shrine restorations, though not technically arts projects, fostered this "sense of kinship" with the past whereby local communities pursued the preservation of community landmarks while positioning the federal government as the benefactor making it possible. Consequently, the historic shrines were not homogenous in character but heterogenous, and reflected a breadth of actors, preservation techniques, and political and economic motivations. The restoration projects, then, should be viewed as experiments in community efforts to strengthen and articulate local cultural identities through promoting historical tourism within the context of a federally funded and politically driven cultural agenda. Similar to the other historical efforts of the New Deal, the shrine restorations illustrate attempts to both diversify public understandings of the nation's history and to reify powerful and often exclusionary historical narratives that constrained rather than expanded what it meant to be American. Much like all facets of Roosevelt's New Deal, we see progressive and conservative strains within the politics of preservation. The next four chapters explore in detail the historical myths, political agendas, economic needs, and cultural anxieties surrounding the execution of four WPA and NYA historic restorations across the United States.

CHAPTER 2

The "Compelling Romance" of the Old South

The Dock Street Theatre, Charleston, South Carolina

> Charleston, S.C., is mindful of her inheritance. Among her historic buildings, impregnated with the spirit of the old south that has gone with the wind, are the Planters Hotel and the Dock Street Theatre. . . . They rise again, reconstructed faithfully by skillful hands. Charleston's inheritance is preserved—not only for Charleston, but for an America thoughtful of her traditions.
>
> Robert Armstrong Andrews, state director of the
> South Carolina Federal Art Project, February 1937

On November 26, 1937, five hundred audience members enjoyed the first performance in more than two centuries within the walls of the Dock Street Theatre in Charleston, South Carolina. The city's Little Theatre acting troupe, the Footlight Players, reenacted *The Recruiting Officer*, a raucous, eighteenth-century Restoration comedy that had opened the original colonial playhouse on February 12, 1736. During those intervening two hundred years, the space that housed the colonial establishment underwent a number of significant physical changes, with the modest theater disappearing entirely to be replaced by a larger set of conjoined buildings which served different functions over time. The Dock Street Theatre had succumbed to fire within its first few decades of operation, and a roaring antebellum hostelry, the Planters' Hotel, stood in its place for much of the 1800s. After the Civil War, the resort fell into disrepair, a state in which the site remained until a new and embellished version of the Dock Street Theatre came to life as a New Deal work relief project in the mid-1930s. With political support from the city of Charleston and funding first from the Federal Emergency Relief Administration

FIGURE 2: Dock Street Theatre (Planters' Hotel) facade, c. 1937. Source: Historic American Buildings Survey SC-467, Library of Congress Prints and Photographs Division.

(FERA) and then the Works Progress Administration (WPA), Charlestonians witnessed the rebirth of a long gone but never forgotten site of colonial prosperity and gaiety.[1]

Opening night of the Dock Street Theatre in 1937 signified the imaginative transformation of the dilapidated Planters' Hotel into an architectural gem for white Charlestonians to celebrate during the bleak years of the Depression. Although the exterior walls of the old hostelry looked the same, albeit given a facelift, its interior underwent a radical alteration as the architects repurposed the hotel spaces into a modern performance venue, inspired by the site's former famed theater. Locals recognized the Dock Street Theatre as a potent symbol of Charleston's prominence in the colonial period as a center of architectural expertise, furniture production, and industries tied to the slave trade and plantation slavery. In deciding to build anew the theater, elite white stakeholders could celebrate the city's historic character and its former cultural supremacy. By the twentieth century, Charleston's cultural arbiters—members of hereditary organizations, old planter-class families, and prominent local artists, politicians, and businesspeople—actively advocated for the preservation of the buildings and streetscapes of the prosperous colonial and antebellum eras as a way to cope with a difficult present and reassert their social and political power diminished by the Civil War.[2] By developing historical tourism that erased more recent and diverse histories of the Planters' Hotel, prominent white actors used New Deal funds to manufacture a powerful image of Charleston's glorified past and reclaim the city's position as a regional center of art.

The resurrection of the Dock Street Theatre during the Depression fit nicely into the portfolio of the Charleston Renaissance, a term used to describe the outpouring of artistic and literary work in the 1920s and 1930s that celebrated local cultural achievements and reflected a profound appreciation of the city's colonial and antebellum history, particularly its architectural heritage.[3] Not coincidentally, the local preservation community's underlying objective to safeguard pre–Civil War structures corresponded with the popularity of this cultural movement. Efforts to preserve the city's early architectural identity targeted the most historic section of the peninsula: the southernmost area, framed by the Ashley River to the west and the Cooper River to the east. The site of the original Dock Street Theatre of the 1700s, located at the intersection of Church and Queen Streets, occupied prime historic real estate in this area. Yet, the extant old Planters' Hotel that stood there was a mixed-race space that became home to both Black and white occupants in the first couple of decades of the twentieth century. To white Charlestonians in the 1930s, this location was the perfect site to materialize an idealized image of the historic city through preservation activity. The restoration's sponsors hoped the theater would project a sanitized portrait of Old Charleston that promoted artistic and architectural excellence and downplayed or willfully ignored racial and class tensions. Thus, local leaders utilized federal monies to boost local tourism based on whitewashed history and displace Black renters who occupied the buildings transformed into the new Dock Street Theatre.

At the same time that this conservative ideology motivated the project, the theater's revival became an important story in the larger narrative of southern progressivism in the New Deal era. While the desire to construct a romanticized version of Charleston reflected the old planter class's attachment to a privileged past built on a slave economy, the preservation endeavor also advanced the city's Democratic agenda to revitalize the struggling port town by using newly available federal funds to update entertainment infrastructure. Upon completion of the new Dock Street Theatre, Douglas D. Ellington, the federal architectural consultant who oversaw the relief project, praised the building's potential to reestablish Charleston as an exciting and lucrative hub of American, not just local or southern, culture that would attract the nation's attention: "[The theatre is] ready to become an active instrument in the public life. . . . Operated in a full sense of idealistic obligation, it could become an instrument of more than local satisfaction, [it] could also be of national value and importance. It is not too extravagant to imagine that an actual cultural renaissance might have founding from within its walls. The building is not merely a theatre, but the planning and arrangement is such that it stands ready to function broadly as a cultural and artistic heart of the city."[4] Charleston's mayor Democrat Burnet Maybank embraced the idea of the theater as a cultural focal point around which to build a historical tourism industry and used the federal project to push forward a modern partisan program.

The young Maybank enjoyed the political and personal support of leading Democrats in Washington, including James Byrnes, South Carolina senator and President Roosevelt's close advisor, and Harry L. Hopkins, WPA national director. As the Roosevelt administration allocated large sums of New Deal money and policy attention to improving the modern economy of southern states, Maybank strategically capitalized on these friendships and new federal support that encouraged the expansion of local governments' modernization efforts, especially with regard to the development of public infrastructure. Throughout his tenure as mayor, he proposed projects that would improve Charleston's economic and cultural arenas as well as garner himself political power as he prepared for higher positions within the national Democratic Party. Undoubtedly, the mayor's ambitious civic projects, including the Dock Street Theatre, found success not only because he courted Charleston's patrician families but because he earned the necessary support of party leaders.

Charleston's impetus to restore the Dock Street Theatre, then, was double facing. In its attempt to recover the past to improve the present, the theater's return to life represented aspirations for contemporary and future Charleston to be *the* historical and cultural tourist destination of the South. The city, under Mayor Maybank, eagerly utilized newly accessible work relief funds that extended federal reach into the state and local government, a concession to New Deal federalism. At the same time, the project still perpetuated the region's characteristic political and social conservatism. The preservation decisions made by the historical actors directing the project illustrated this ambitious agenda to both consolidate white power in the historic built environment and push Charleston to embrace some degree of change, such as accepting help from the federal government.

Reflective of the fact that the design goal guiding the theater project was not to faithfully and accurately reconstruct a modest colonial establishment but to shape the city's historical image in ways that would dominate the burgeoning and profitable tourism industry, the politicians and architects in charge made the most creative choices possible in terms of preservation methodology. Because the theater no longer existed, any iteration of a playhouse unquestionably would be a new build. Accordingly, the architects concocted an imaginative plan: they created a site with the facade of one building—the still-standing nineteenth-century Planters' Hotel—behind which they placed a newly designed theater, inspired by a colonial venue that long since had vanished. Moreover, they outfitted the space with authentic historic architectural pieces salvaged from a nearby early nineteenth-century Federal-style residence, the renowned Radcliffe-King House. In other words, a historic exterior clothed a fabricated interior featuring "real" artifacts. These intentional design choices revealed the ultimate purpose of the preservation endeavor from an architectural standpoint: to produce a believable but essentially fantastical reproduction of the eighteenth-century Dock Street Theatre that ingeniously combined actual historic materials with new construction.

For proud, history-loving, white Charlestonians in the early twentieth century, recreating the extinct colonial theater restored their sense of cultural primacy that the Civil War had destroyed. The word "restoration," used here erroneously to describe a rehabilitation and new construction, cuts to the heart of the project to reclaim the authority elite stakeholders once held over all aspects of life in Charleston. For the federal government, which played the important role of financial benefactor, the revived theater presented an opportunity to showcase the wealthiest colonial southern city's cultural refinement, an inspiring reminder of affluence and gentility during a period of deep economic troubles. The federal and local motivations combined to make the reemergence of the Dock Street Theatre in Charleston's physical landscape a creative cultural and political experiment that exemplified the wielding of the built environment to reinforce prevailing conservative mindsets while also promoting community revitalization through the development of white historical tourism. Although inherently looking to the past, as all preservation projects do, the Dock Street Theatre's ceremonial return to the peninsula embodied a powerful, forward-looking ambition to use historic architecture as a means to reassert control. In other words, the "restored" theater called on Charleston's historic sense of place to serve both contemporary and future generations, becoming a tangible form of political and cultural agency.

Charleston's Decline Spurs Preservation Efforts

The first Dock Street Theatre, constructed in 1736, sat on the southwest corner of Church and Queen Streets.[5] The name of the original thoroughfare called Dock Street changed to Queen Street in 1734, and although the first advertisement in the *South-Carolina Gazette* referred to the newly built structure as the "New Theatre in Dock Street," subsequent mid-eighteenth-century notices called it the "Theater in Queen-Street," "Play House," or simply "the Theatre."[6] The modest playhouse enjoyed a robust first spring season, although it attracted a less elite clientele than comparative colonial sites of public entertainment in town, such as McCrady's Tavern and the dining and entertainment venue Longroom. Despite its early success, the theater faced difficulties in subsequent years, changing ownership within six months and again several more times before burning to the ground in 1754.[7] In 1809, Alexander Calder and his wife, proprietors of the nearby Planters' Hotel, purchased property from Major John Ward on the corner of Church and Queen Streets where the Dock Street Theatre had stood in the previous century. The Calders relocated their popular hostelry less than a block to the east of the site of the old theater, and remodeled and enlarged the Planters' Hotel several times over the next couple of decades.[8]

The bustling activity of the Planters' Hotel in the mid-nineteenth century contributed to what Eola Willis, amateur theater historian of Charleston, dubbed "the Jolly Corner" in her 1924 book *The Charleston Stage in the XVIII Century*. As a founding member of the Society for the Preservation of Old Dwellings (SPOD) and historian of the Poetry Society of South Carolina, Willis was deeply invested in celebrating the city's ancestral heritage. While her magnum opus on the city's early theater history should be approached as the work of an enthusiast rather than a professional, it helped perpetuate the myth of Old Charleston as a flourishing, sophisticated colonial capital and the heart of the antebellum South.[9] At the intersection where the Planters' Hotel reigned, Willis wrote, "gentlemen of the old regime met to discuss horses, politics, and the events of the times," while sipping the popular rum concoction Planter's Punch, supposedly invented at the establishment.[10] From year to year, wealthy planter families from the Carolina Upcountry and neighboring regions lodged at the busy hostelry for several weeks during the spring social season, and stagecoaches from Savannah and cities farther west started their return journeys from the hotel.[11] It is likely that the property received a facelift during these prosperous years. In 1852, John C. O'Hanlon assumed ownership of the hotel and presumably adorned the place with its notable facade: rusticated brownstone columns, cast-iron balcony, and decorative mahogany brackets (see fig. 2).[12] In the next decade, the Planters' Hotel took on added import when it served as a gathering hub and informal communications center during the Civil War.[13]

The Planters' Hotel survived the destructive conflict but never recovered its antebellum fame. By the 1880s, the handsome and once lively establishment had become neglected tenement housing for poor Black Americans, and the complex continued to deteriorate in the early decades of the twentieth century. In 1913, the secretary and chief case worker for the Associated Charities Society of Charleston, an organization composed solely of white membership, described an encounter in 1913 with a transient single mother, a Mrs. Phillips, driven to begging who was temporarily residing at the old Planters'

Hotel. The case worker did not record Phillips's race, but the society aided both white and Black residents, as well as immigrants, and the area was home to racially diverse occupants.[14] Nearby Cabbage Row (or "Catfish Row"), of famed local author Dubose Heyward's novel *Porgy and Bess* (1925), suffered the same fate as the Planters' Hotel. In 1922, white neighbors upset with the changing racial demographics of the area successfully petitioned Charleston's city council to evict Black tenants then occupying Cabbage Row, leading to its vacancy in the 1920s.[15]

The degradation of the physical state and evolution from white to Black occupants of these two once-elite sites reflected the broader context of changing political and racial conditions following the Civil War. Increasingly, legal segregation manifested in public and commercial spaces during the Jim Crow period, but private spaces also underwent shifting demographics.[16] Some areas of Charleston's historic core experienced disinvestment of the care and capital required for improvements, a process represented by the worsening condition of the former Planters' Hotel. Indeed, the once celebrated locale had altered so notably that in the 1930s, the editor of the *Charleston Evening Post* and chair of Charleston's Board of Architectural Review described the structure as "a gaunt and sometimes dangerous relic."[17] The "Jolly Corner" of the city's renowned colonial theater and later premier antebellum hostelry had become, in local opinion, an eyesore. Even more, it challenged the idea that the geographic heart of colonial Charleston historically was, and remained, an exclusively white space.

The material decline of the Planters' Hotel was not altogether surprising, as the city steadily witnessed the loss of historic fabric amid general economic hardships in the decades after the Civil War and well into the twentieth century. Charleston faced the challenges of a sluggish economy hit hard by the collapse of the cotton market, and a demoralized white society unable to foster a thriving business class to compete with other growing southern cities such as Atlanta and Nashville. Moreover, by the 1890s, the growth of railroads in the South had undermined Charleston's historic role as the chief port on the southern Atlantic Coast.[18] Partly in resistance to unwelcome changes caused by the profound collapse of the South's social and economic system built on slavery, the city's wealthy white families nurtured a strongly conservative ethos. In the words of one historian, Charleston's commitment to rigid societal standards solidified its role as the "social arbiter" of the state even when other aspects of life were overturned.[19]

Accordingly, planter-class families institutionalized their conservatism in the early twentieth century through the founding of local organizations that dictated the city's cultural scene for decades. For example, the Charleston Art Commission formed in 1910 as a challenge to the nationwide City Beautiful urban planning movement, focusing instead on maintaining the "city historic"; the Poetry Society of South Carolina launched in 1920; and the highly exclusive—and peculiarly white only—Society for the Preservation of Negro Spirituals emerged in 1922 to preserve the Gullah folk music of the Lowcountry plantations.[20] These organizations and their limited memberships, which to an extent embraced women, established the portfolio of Charleston's cultural goods and demonstrated the powerful legacy of Progressive-era ideals surrounding order, race, beauty, and sanitation.[21]

The conservative economic and cultural attitude of Charleston's white elite shaped the social and physical geography of the city, and vice versa. After the Civil War, old families refused to leave their decaying mansions, a prideful act that produced a "museumlike quality" to the historic environs and gave credence to the local saying, "Too poor to paint, too proud to whitewash." As historian Don H. Doyle has argued, the signs of genteel poverty became "proud badges of a déclassé aristocracy who refused to answer the siren call of the New South." Unlike in some southern cities, such as Atlanta and Mobile, where wealthy families left their urban mansions for the nascent suburbs, Charlestonians held on to their downtown historic homes as the last vestiges and symbols of their former power.[22] These families most likely viewed the transition of once elite white spaces like the Planters' Hotel to transient housing for people of color as an upsetting phenomenon that further materialized physically the loss of their control over social spaces. They lamented the destruction of the beautiful historic architecture of their city, while at the same time selling old furniture, silver, fireplaces, ironwork, and other family heirlooms to antique dealers from the North. The desire to protect Charleston's architectural heritage from the greedy hands of outsiders—especially northerners—galvanized some prominent residents to form the city's first formal preservation organization, the SPOD, in 1920.

Spearheaded by the SPOD, preservation activity in Charleston during the interwar period prioritized colonial and antebellum structures of white residents that carried a history of prestige and power. Because of its location on a busy intersection in the much-visited historic area, the rundown Planters' Hotel attracted the attention of these early preservationists. Under the leadership of Susan Pringle Frost, the SPOD saved the Planters' Hotel when the city scheduled its demolition in 1918. Frost successfully persuaded Mayor Thomas P. Stoney and the city council to seal the four conjoined buildings that comprised the corner hotel until a time when they could be restored to their former antebellum splendor. The "sealing" of the attached buildings protected the exterior walls, maintained the condition of the interior, and, importantly, prevented their demolition. Additionally, the act barred wealthy northerners from acquiring the Planters' Hotel's valuable ironwork, woodwork, and plaster.[23] Rather than committing time and money to a preservation campaign to restore the property, essentially by then a commercial space turned residence for mostly poor Black tenants, the SPOD chose to focus its efforts on rescuing the homes of South Carolina's leading families: the threatened Joseph Manigault House and the Heyward-Washington House.[24] The SPOD's choice illustrated two important national trends of the early twentieth-century preservation movement: attention to elite structures over vernacular ones—in this case, upper-status domestic dwellings over former commercial sites—and the valuing of buildings firmly associated with whites rather than racial or ethnic minorities.

The city council codified this limited outlook of the city's preservation movement, prescribed by prevalent gendered and racial codes, when Chaleston created a planning and zoning commission. The newly minted commission enacted the country's first historic zoning ordinance in 1931, which designated a twenty-three-block area in the tip of the Charleston peninsula as the "Old and Historic District," sending a clear message that within its perimeter racial hierachies and conservative values would be preserved intentionally in the built landscape. While protecting domestic architecture, the zoning ordinance allowed for the commercialization of Church Street as attractions such as Cabbage Row, the Heyward-Washington House, the Porgy Book Shop, and antique stores and coffeehouses drew growing numbers of tourists to admire Charleston's historic charm.[25]

Historical tourism was not the only factor drawing attention to the city's built environment. Various New Deal programs also effected radical change in the racial dynamics of the city's urban landscape. In the fall of 1935, Charleston received a $1.1 million federal grant for the clearance of Black neighborhoods, beginning a pattern of government-sponsored gentrification that later would reemerge in the construction of highways and urban renewal programs of the mid-twentieth century. Throughout the late 1930s, Black residents living near the historic district and denied political voice through Jim Crow laws were displaced and relocated to public housing projects up the Charleston Neck, farther from the downtown area.[26] The Santee-Cooper hydroelectric project, the largest New Deal initiative in the state and partially funded by the Public Works Administration and the South Carolina Public Service Authority, disproportionally affected Black people. Of the 901 families who suffered discolation from the Santee-Cooper project, 800 were Black.[27] The forced movement away from the city's historic core helped escalate a significant population shift initiated by the introduction of the U.S. Navy to the Charleston peninsula at the turn of the twentieth century: what was once a Black-majority region became 55 percent white by 1930, and the balance remained that way for the next two decades.[28] In addition to demographic changes, these discrimatory housing measures effectively minimized the "casual mixing" of residents, whereby Black occupants lived in the former quarters of enslaved people or servants at the back of white-owned townhouses, which was both a legacy of urban slavery and a result of a population almost equally divided between the two races.[29] The strengthening local preservation impulse to protect historic structures associated with white pre–Civil War history thus combined with the power and money of new federal programs and discriminatory legislation to engender a lasting shift in longstanding living arrangements.

It is important to note that some local preservationists did condemn various New Deal efforts which threatened the historic built environment and residential patterns, expressing disapproval of the construction of low-cost housing and slum clearance. Their opposition, however, generally was born out of reverence for historic architecture representative of the region's plantation past and objections to environmental destruction rather than respect for the lived experiences of marginalized Charlestonians. Thus, local objections to the hydroelectric project exhibited the same attachment to the privileged racial and social orders that fueled support of the simultaneous Dock Street Theatre project.[30] As historian Jack Irby Hayes Jr. has concluded best, with New Deal funding dramatically changing the peninsula's physical landscape, "Charleston preservationists were ambivalent."[31] According to the 1941 WPA state guide to South Carolina, white residents "with a love of the unusual" paternalistically reclaimed city spaces historically occupied by Black tenants with the goal to renovate, thereby reconfiguring the urban geography with little concern for the consequences for the former inhabitants.[32] Upon witnessing the restoration work occurring on popular streets in the Old and

Historic District, including Tradd, lower East Bay, Church, and Stolls Alley, local artist Elizabeth O'Neill Verner praised "what can be done if cleaning up infested neighborhoods and turning our liabilities into assets."[33] The language evident here highlighted the purposeful elimination of unwanted residents occupying desirable, if rundown, historic spaces. This early gentrification made downtown a white-dominated space, which it never had been historically. The process only intensified in the coming decades as tourism increasingly centered around Charleston's image as a living historic *white* city, and expanded federal programs in subsequent decades executed larger-scale and more damaging "slum clearance" coded as urban revitalization.[34]

The Dock Street Theatre Becomes a Work Relief Project

The dilapidated and abandoned Planters' Hotel, then, in its prime location in downtown Charleston, presented both a problem and an opportunity as former tenement housing for Black residents in a historically significant area. A proposal from the political elite and New Deal money provided the solution. It is possible that Elizabeth Maybank, a prominent member of the Junior League and wife of the mayor, first proposed the idea to restore the Dock Street Theatre as a city project in early 1934.[35] Mayor Maybank most likely then suggested to the Charleston Art Commission the transformation of the old hotel back to the Dock Street Theatre that once occupied the corner as a potential federal project that spring or summer. Afterward, he called a special meeting of key local figures to discuss the proposal in mid-October. In addition to regular members of the commission, attendees at the meeting included New Deal officials Edmund P. Grice, the Charleston County administrator of the FERA, and Douglas D. Ellington, the FERA's architectural consultant. The committee agreed that the area surrounding the famed St. Philip's Episcopal Church on Church Street near its intersection with Queen Street—exactly where the Planters' Hotel stood, as seen in figure 3—was the most suitable for architectural restoration because it included many historic landmarks that had fallen into disrepair, including the Powder Magazine along with the once-famed hostelry.[36]

After the committee selected to transform the Planters' Hotel into the Dock Street Theatre, Mayor Maybank corresponded with Harry Hopkins, a close personal friend and, at the time, the federal director of the FERA.[37]

FIGURE 3: View of Charleston's Church Street facing north. The Dock Street Theatre is on the left and St. Philip's Church is in the center. Source: Frances Benjamin Johnston, Carnegie Survey of the Architecture of the South, 1936–37, Library of Congress, Prints and Photographs Division.

According to Albert Simons, the Charleston architect whose firm would lead the restoration project, Hopkins "was immediately attracted by the plan, since it eminently fulfilled the government's desire to underwrite projects which would provide work for the unemployed as well as be in themselves constructive and worthwhile."[38] Moreover, the Roosevelt administration was eager to allocate funds to projects that would revitalize the South; a historic preservation project, with its potential to provide jobs for the unemployed and boost the local tourist economy, easily satisfied this mandate.[39] Throughout Maybank's tenure in office and Hopkins's leadership first of the FERA and later the WPA, the two developed a mutually beneficial partnership that strengthened Democratic politics and the reception of New Deal initiatives in Charleston. Maybank shrewdly used his close ties with influential New Dealers to secure resources for grassroots ventures. While mayor, he advantageously sat on the board of three New Deal agencies: the Public Works Administration, the State Board of Bank Control, and the South Carolina Public Service Authority, which oversaw construction of the Santee-Cooper hydroelectric dam.[40] Maybank's valuable friendships and his consistent political backing of the Roosevelt administration undoubtedly helped produce a favorable attitude toward the proposed preservation project at all government levels, and ensured that the two-and-a-half-year endeavor secured sufficient funding to see it to completion.

In early February 1935, Maybank announced that upon the city council's recent agreement to purchase the Old Planters' Hotel, the Dock Street Theatre officially became a FERA project.[41] "Two birds are being killed with one stone," the *Charleston News and Courier* jubilantly declared. First, after years of talk about its restoration, the Planters' Hotel would be "transformed from an eyesore to a place of beauty." Second, the city had found a site for a new theater, which it had been hoping to construct for some time.[42] Once on the FERA docket, the Ways and Means Committee, headed by Charleston County FERA administrator Edward P. Grice, appropriated $10,000 for a sixty-day period of initial survey work, which began on February 12. The next month, after praising Maybank's efforts to get the work underway, Hopkins remarked that the FERA "will pay any reasonable amount for labor and materials" when asked about the expected cost of the restoration.[43]

By early May, Washington officials had given final approval to the plans of FERA architect Douglas Ellington, who then met with relief administrator Grice and Mayor Maybank several times to discuss next steps. By June, ten workers and six carpenters were engaged on the worksite and the FERA already had expended $159,000. A press release of the newly created WPA issued that month promised that the project "begun last fall by a group of Charleston citizens" would provide employment for "about 80 workers from the relief rolls," including both blue- and white-collar laborers.[44] Work that summer included the "demolition of destroyed portions of interior"; "removal of all debris," which was mostly rotted wood and plaster; shoring of brickwork; preparation of structural, mechanical, and electrical drawings; and "cataloguing . . . of any materials of value in restoration."[45] While excavating, workers discovered marks in the roof of the Planters' Hotel from a cannon shell, perhaps fired during the bombardment of Charleston in 1863, as well as old coins and fragments of broken china.[46]

When the Dock Street Theatre project transitioned from the FERA to the WPA during the fall of 1935, there was confusion about whether the restoration was associated with the Federal Theatre Project (FTP). Multiple Charlestonians, including theater historian Eola Willis, the president of the Footlight Players, and union stage workers, erroneously believed that the FTP would administer the project. Hallie Flanagan, national director of the FTP, also indicated uncertainty early on about whether the theater would be added to her program. She expressed interest in the Charleston project, and her assistant even reached out to Helen Smith Whaley, wife of former Charleston mayor Goodwyn Rhett, to ask for information about the project from someone keyed into local affairs. As Flanagan's assistant explained, "We hope under this new program of Federal theatre projects to establish certain historical theatres in America," and Charleston "seems to be a logical place for one of these units."[47]

The president of the Footlight Players, the acting troupe that would first perform in the theater, also believed the project to be aligned with the newly established FTP. J. P. Frost wrote to Flanagan to offer the Charleston organization's cooperation, but to the director of the Town Theatre in the state's capital of Columbia he called the FTP "a rather disturbing project." He wrote, "We are regarding the Dock Street Theatre restoration with very interested and greedy eyes, and are hoping that this Federal Drama Project will not interfere with our ambitions." Flanagan's response to Frost was positive, but Frost remained skeptical about whether there were enough "theater people on relief rolls," nor did he think the Footlight Players "could 'absorb' a group of unemployed actors."[48] Frost's concern over potential conflict between his troupe and an FTP unit was framed in terms of labor issues, but it also demonstrates reluctance to cede control and use of local institutions to a federal program, especially one endorsing progressive racial ideas through the establishment of Negro Units. Flanagan's support of interracial theater and her interest in Russian arts would have directly challenged the status quo of segregation and Americanism that the Dock Street Theatre's local boosters so clearly wanted to protect.

Flanagan forwarded Frost's letters to Frederick Koch, who had been appointed the FTP's regional director for North Carolina, South Carolina, and Virginia, but by the end of November, the question of whether the FTP was to be involved with the Dock Street Theatre was definitively answered in the negative. When a theater director wrote to Flanagan asking if he could be assigned the directorship of the new venue, she replied that while she maintained interest in the Charleston project, she had "been informed that this project is under a different department."[49] Indeed, the Federal Art Project of Federal One oversaw set production, but the relief project remained under the WPA's Division of Engineering and Construction. The Dock Street Theatre had no formal connection to the FTP, although Flanagan continued to offer suggestions for its management and visited the site in 1938.

While Ellington devised the provisional plan for the theater, which he reportedly described as the largest restoration ever undertaken as an emergency measure by the federal government, Charleston-based architects Albert Simons and Samuel Lapham began to marshal the project locally.[50] Working in consultation with Ellington, the duo drew detailed architectural plans and oversaw the day-to-day work, playing a much larger role in articulating the theater's design than the federal architect.[51] As native sons, Simons and Lapham dedicated their careers to safeguarding Charleston's architectural heritage. Simons served as a member of the city's Planning and Zoning Commission, president of the Carolina Art Association, and instructor of architecture at the College of Charleston, where Lapham also taught. Both architects became involved in the Historic American Buildings Survey (HABS) when it was established in 1933 by the Civil Works Administration. Secretary of the Interior Harold L. Ickes appointed Lapham to serve as HABS district administrator in South Carolina, and Simons sat on the National Advisory Board overseeing the federal program.[52] The two architects were uniquely qualified to direct the theater project both as professionals and as bona fide Charlestonians, a claim that mattered in a place where family name carried social weight. Once on board, Simons and Lapham immediately embraced the plan to recreate the appearance and feel of Old Charleston through manipulating the historic built environment.

As keen architectural historians and practicing architects, Simons and Lapham did their due diligence and sought original plans of the Dock Street Theatre before designing the new playhouse, but their research proved fairly fruitless.[53] Despite Hopkins's public assurance that Ellington already had "succeeded in obtaining a very good idea as to what the original theater looked like," all three architects were unable to find floor plans or detailed accounts of the 1736 venue. A 1739 map of Charleston indicated the location and relative size of the first Dock Street Theatre, but colonial newspapers contained only short descriptions or brief mentions of the space.[54] Without building plans or even a sufficient description of its interior, Simons and Lapham could embellish the smaller, simpler eighteenth-century structure in a fashion that suited their vision of colonial grandeur. They decided to base their design on the style of quasi-contemporary English playhouses, particularly those in London. With help from Library of Congress staff, the architects found a reproduction of Christopher Wren's design for London's Drury Lane Theatre, an English Restoration–style structure built in 1674.[55] The London theater was much larger and grander than Charleston's colonial playhouse and predated it by over sixty years, but it matched the architects' vision of a thriving, culturally rich port city.

During this early phase of the project, Harry Hopkins announced to the city that "our tentative research points very convincingly to the probability that the old theatre, which has been supposed to have been the third building of its kind in the United States, was actually the first. This adds to the historical importance of the reconstruction project that has been launched."[56] Indeed, the WPA press release announcing the project described it as "rebuilding America's first theater," backed by "various old documents and the yellowing files of the *South Carolina Gazette* [that] show that Charleston can properly assert its priority."[57] The claim of Charleston having the first purpose-built theater, however, was not uncontested; both Kenneth Chorley, president of Colonial Williamsburg, and Harold H. Shurtleff, its director of research and records, argued that the theater of Virginia's colonial capital dated to 1716, predating the Dock Street Theatre by twenty years.[58] Despite their efforts to convince both Ickes and Hopkins of their conclusion, the WPA continued to advertise the Dock Street Theatre as the first built in the nation. Regardless, Charleston could boast an earlier theater history than most American cities, a fact remarked upon in a *New York Times* article that described the Dock Street Theatre as fifteen years older than the Nassau Street Theatre in New York City and thirty years older than the Southwark Theatre in Philadelphia.[59] Charleston led the way in entertainment for burgeoning cities in the Northeast, earning its title as "the grandfather of America's great theatre industry."[60]

Charleston's assertion of being the first to offer theater as entertainment for its residents became an important promotional feature of the restoration project because it reinforced an image of Charleston's role as the premier cultural center of the South in the eighteenth century. In an article written for the Associated Press, managing editor of the *Charleston News and Courier* Thomas P. Lesesne expressed pride in Charleston's association with refined entertainment when other southern cities were in their cultural infancy; at the time the Dock Street Theatre opened its doors, for example, "Savannah was in its swaddling clothes."[61] The *Atlanta Constitution* descriptively offered that when *The Recruiting Officer* first premiered in Charleston in 1736, "Indians and roughly clad backwoodsmen" and "the things of a wilderness primeval" were still common to the outposts of the British Empire.[62] While other southern cities remained largely unsettled and provincial, Charleston had emerged as the leading colonial city in sophisticated cultural affairs.

Simons and Lapham had their work as restoration architects cut out for them in order to invent this illustrious historic Charleston promoted by politicians and the press. While the restored theater needed to establish a visual and experiential connection to the past, the new space also needed to be functional and not merely an architectural showpiece for Charlestonians to exhibit to tourists. Simons expressed the opinion that it was imperative the Dock Street Theatre become "a living part of the community" and not "a museum piece, exquisite, but useless."[63] The task to build a new theater behind the exterior walls of the Planters' Hotel presented both creative and technical challenges since what was called a "restoration" was really a rehabilitation of the extant structure and new construction of the long-vanished theater. The 1930s project area included the four conjoined brick buildings that comprised the Planters' Hotel, extending 155 feet south from Queen Street and 120 feet west of Church Street. Three of the four buildings faced Church Street and had three stories: the first sat on the corner of the intersection, the second was the middle brownstone structure with more elaborate decoration that served as the hotel entrance, and the third occupied the space just south of the main hotel building. The fourth building, taller than the others at four stories, faced Queen Street. According to Simons, by 1935 the four buildings "were all but shelled. As though gutted by fire, virtually all the interiors were completely gone or rotted beyond repair."[64] Effectively, and as much admitted by the architects, any interior "restoration" work would be, in truth, highly educated and thoughtful guesswork.

Diving into the task at hand, the architects creatively reconfigured the existing buildings to produce a theatrical space inspired by elegant colonial theaters that would also account for modern convenience and entertainment needs. They designed a two-story theater to fill the area formed by the "L" of the four attached buildings, making it invisible from the street. The audience would sit facing Queen Street with the stage built to abut the four-story building to the north. The total project area, thus, created a rectangular space, with the theater occupying a much larger footprint than its colonial predecessor. The theater and secondary rooms dominated much of the main floor, but plans for the building in the southeast corner of the project included a restaurant and dining room. The architects intended to transform the southwest portion of the rectangle into an outdoor courtyard with a fountain and space for dining. Offices, a balcony, and foyer would occupy the second floor of the rehabilitated buildings, while the third floor was to be divided into eight apartments for white occupants.[65]

Unsurprisingly, the architects paid particular attention to plans for the main attraction: the theater itself. Simons and Lapham believed the project's credibility rested largely on their adherence to principles of colonial theater architecture, even if the construction and style did not actually mimic Charleston's 1736 playhouse. As English Restoration–style theaters of the seventeenth century typically featured a proscenium arch that framed the stage and bench seating, Simons incorporated five hundred tilted seats attached to a bench-style back in the Dock Street Theatre's design. The architects also built thirteen viewing boxes seating eight persons each to flank three sides of the theater and a gallery at the back of the main space (fig. 4).[66] The new theater could accommodate six hundred people, which was probably at least three times the capacity of the original structure.

Besides drawing on London playhouses for inspiration, Simons honored Charleston's own built heritage in the interior decor. He mimicked nearby St. Michael's Church in the coved ceiling of the theater and copied the British coat of arms above the stage from St. James Church in Goose Creek, an area north of Charleston.[67] Figure 4 illustrates the stately decorative elements designed for the new theater that fashioned an environment in which the audience would "have the illusion of sitting in an 18th century playhouse," since the genuine experience was denied them.[68] Georgian-inspired woodwork of black cypress gleamed from an applied mixture of vinegar and iron filings, drapery decorated the viewing boxes and served in place of doors over the entryways leading from the lobby to the theater, chandeliers hung from the ceiling, brackets along the paneled walls encased electric candles, and a black metal ring suspended by chains held candle lights that hung in front of the stage.[69]

FIGURE 4: Interior view of the Dock Street Theatre's auditorium, featuring bench seating and a proscenium arch above the stage. Source: Frances Benjamin Johnston, Carnegie Survey of the Architecture of the South, c. 1936–37, Library of Congress Prints and Photographs Division.

While the new theater represented the eighteenth century only symbolically, the extant buildings utilized for the project remained physical, aging legacies of the nineteenth century. The architects aimed to protect the patina of the Planters' Hotel's facade on Church Street, the image generally associated with the project and visible to passersby. Simons and Lapham made sure to preserve the most distinguishing features of the old structure: the original cast-iron balcony in a morning-glory pattern and sandstone entrance columns with rare carved mahogany cornices from Barbados (as seen in fig. 2). Without compromising the integrity of the historic material, the architects repaired the brickwork and the balcony, installed new window sashes and frames, and applied a thin color wash to the repaired walls to duplicate "the soft rose of the old stucco." In addition, they reinforced the exterior walls and foundations, and rebuilt the roof, floors, and partitions with mostly steel and concrete, modern materials that they hid from sight.[70]

Architects Ellington, Simons, and Lapham later described in detail the preservation philosophy that guided their creative approach to the surviving historic structures in the January 1938 issue of *Architectural Record*: "The technique of restoration in Charleston differs substantially from that in vogue elsewhere in that it is 'freer' and tends to preserve, externally at least, the cumulative effects of age and use."[71] Their explanation illustrates that the architects did not feel bound by the many limitations of their site; they embraced the unconventional nature of the project to merge an imagined theater with an extant structure. The nineteenth-century facade of the Planters' Hotel, after all, irrefutably offered tangible history and lent credibility to the project as a *restoration* of a genuine historic space rather than merely new construction. Yet, the architect's "freer" approach, as they acknowledged, also contrasted significantly with that of contemporary expert restoration architects performing preservation work elsewhere, as seen in the punctilious attention to the Whitfield House's architectural history discussed in chapter 3. Furthermore, in choosing to protect the hotel facade, the Charleston architects kept intact the building's character when occupied by Black renters, unintentionally preserving an otherwise neglected historical narrative in the New Deal project.

Salvaged Goods: Old Is Better Than New

To further enhance the ambience of Old Charleston in a space containing mostly new materials, Simons and Lapham used architectural elements salvaged from a nearby nineteenth-century mansion, the Radcliffe-King House, in the theater's interior. The relocated ornamentation played a crucial role in creating the illusion that the new Dock Street Theatre maintained an unbroken connection with the Charleston of yesteryear despite the damaging fire that ravaged the colonial playhouse, the rise and fall of the antebellum Planters' Hotel, and the affronting deterioration that was a consequence of tenement housing. The Radcliffe-King pieces reminded visitors of Charleston's former glory and boosted the potential prosperity of the theater by embellishing it with artifacts from a notable and period-appropriate historic home.

The Radcliffe-King House was a Federal-style mansion built in the first decade of the 1800s by the wealthy merchant George Radcliffe. It sat on the corner of Meeting and George Streets near the homes of two other prominent Charlestonians, Gabriel Manigault and Thomas Pinckney. In 1824, leading South Carolina jurist Mitchell King bought the estate from the Radcliffe family and turned his home into a center of literary and artistic life. He entertained prestigious guests, including the famous English novelist William Makepeace Thackeray and American novelist and historian William Gilmore Simms.[72] When King died in 1862, his son assumed ownership of the home and later sold the property to the city of Charleston in 1880. The city council then rehabilitated the building for use as the new public high school for boys. When enrollment increased to over five hundred pupils, the city decided that the building no longer met the school's needs and abandoned the property in 1922, leaving it to physically decline much like the Planters' Hotel at the time. The disregard for both once-notable properties is indicative of the poor economic circumstances of the 1920s and also of the preservation community's deprioritizing of commercial, vernacular, and other nonresidential structures.

In the 1930s, however, Simons and Lapham took notice of the unoccupied and threatened mansion-turned-school: here was an authentic relic of Old Charleston, devalued by the city and unworthy of a preservation campaign but with original architecture perfect for the city's new theater. The two architects already had earned themselves the reputation of a preservation-minded firm by recycling materials from buildings endangered by demolition to use in their projects. Simons especially viewed the salvaging of Charleston's architecture as a way to combat the loss of civic and cultural identity that resulted from the tumultuous changes of the previous half-century.[73] When he learned that the Board of Public School Commissioners planned to destroy the former Radcliffe-King mansion, he urged Lapham to ask his father, a city councilman with connections to the schoolboard, to allow their firm to conscript architectural elements from the house's interior.[74] Although Ellington acknowledged that "the cost in removal, cleaning, repairing and reinstallation would be so much greater than the cost of new woodwork," he advocated for funds to be expended to acquire the old materials out of desire rather than need.[75] After "some difficulty about the King Mansion interiors," probably over purchase cost, the schoolboard granted Simons's salvage request.[76] When the board listed the building for sale for $25,000 in 1935, it did so with the condition that the interior woodwork be retained for use in the Dock Street Theatre. The Charleston Museum became the pieces' repository until Simons and Lapham were ready to install them.[77]

The Radcliffe-King elements were Adam style, a European neoclassical decorative style popularized in post-Revolutionary America under the name

FIGURE 5: Salvaged mantelpiece from the Radcliffe-King House, embellished with Corinthian columns, a biblical scene, and floral drapery. Frances Benjamin Johnston, Carnegie Survey of the Architecture of the South, c. 1936–37, Library of Congress Prints and Photographs Division.

Federal style. They included elaborate mantelpieces, Palladian windows, scrolled plasterwork, wainscot, cornices, and mahogany doors. The mantelpieces, relocated to the theater's green room (traditionally a retiring place for actors), were some of the most decorative and impressive salvaged items, featuring Ionic and Corinthian columns, biblical scenes, angelic figures, and draped floral embellishments, as seen in figure 5.[78] Simons described the installation of this repurposed decor as difficult and necessitating exceptional consideration: the pieces' removal, transportation, and placement in the new theater space "represented a special problem calling for the utmost care and skill."[79] The architects' delicate work gained the attention of one local newspaper editor who reverentially termed the installed pieces "Relics Preserved" in the *Charleston Evening Post*. To sympathetically integrate the salvaged architecture into the theater, the "usual construction procedure had to be reversed." Rather than the interior woodwork being designed to fit the space, the openings had to be fitted for the Radcliffe-King pieces "in order that the symmetry, proportion and design of these valuable features of the building might not be marred."[80]

The material culture rescued from the antebellum home created a visual and physical continuity with the past for theatergoers in the late 1930s. The incorporation of these high-style architectural elements from a leading family's former home captured the very essence of Old Charleston, as the city's domestic historic architecture was the most commanding symbol of the conservative and powerful planter class. Moreover, these real historic artifacts of a previous era in an otherwise imagined historical space became an especially important method of upholding the theater's mirage of authenticity. As historian W. Fitzhugh Brundage has argued, historic architecture "made tangible the mythic colonial and antebellum South, allowing visitors to experience firsthand remnants of what was purportedly one of the nation's most elegant and refined societies."[81] In the 1930s, romanticizing the historic South, especially through architecture, emerged as a common trend of popular culture as it reinforced traditional values and offered a creative escape from modern troubles.[82]

Charleston's literary celebrity DuBose Heyward expressed this ability of architecture to preserve and perpetuate cultural values and previous manners and modes of living in his appraisal of the project. He fittingly described the Adam-style woodwork as bringing to the new theater "not only its beauty of plaster and woodwork but its wealth of tradition extending far back into Charleston's past." He continued with an account of the complementary architectural styles combined in the one preservation project:

> Here was no slavish reproduction of a single period, but a bringing together under a single roof of an early eighteenth-century theatre, a group of simple early Charleston dwelling houses, an unmistakable example of the Classic Revival, and the harmonious incorporation therein of interior decoration removed bodily from a Georgian mansion. The harmonizing of these various factors, the ingenuity and taste with which they were merged one into another, and the delightful element of the unexpected which one now encounters in passing from room to room, give this building a character unique in American restoration.[83]

If the rebuilt theater was meant to capture the *spirit* of a 1736 playhouse, the architectural pieces moved from the Radcliffe-King House into the rooms of the rehabilitated Planters' Hotel "uniquely," if not faithfully, put Charleston's prosperous antebellum history on display. Three-dimensional objects can serve as "memory cues, as souvenirs in a quite literal sense," material culture scholar Leora Auslander has contended.[84] While attendees in the 1930s could not walk away after an evening of entertainment with a piece of architecture, the theater itself could become a spatial memory cue recalling a flourishing and white-dominated Charleston. The Radcliffe-King pieces were integral props in painting this immersive scene that rejected reality in favor of reproducing a bygone era.

Finally, into this creative experiment the architects introduced the twentieth century—contemporary Charleston—through modern lighting, sound, and stage equipment brought into the new performance space. The Carolina Art Association boasted that the fifty-six-by-thirty-six-foot stage with a three-story fly loft and projectors for motion pictures was of "the most modern design and far more complete than in any other theatre in the south."[85] The new theater also featured a revolving stage to enable quick scene changes. A WPA relief worker on the job described the switchboard as "of the most modern kind," with trapdoors in the stage floor. Moreover, the fireproof stage had an asbestos curtain with fusible lengths, which, like solder, melted and released the curtain at a certain temperature.[86] According to Heyward, the effect of the architects' sensible and inspired design combined "to an extraordinary degree the atmosphere of the past with the elaborate equipment of the modern theatre." Wrapped in eighteenth- and nineteenth-century decorative garb, the twentieth-century theater of reinforced concrete was built to "withstand the assaults of centuries."[87]

"The Springing of a Memorable Past into This Present New Deal Modernity"

Two months before the theater was set to open its curtains in late November 1937, Douglas D. Ellington told the city's Exchange Club that "Charleston probably is creating the germ which will serve as the nucleus for the national theater movement in America."[88] His words conveyed a tall order, and the project accordingly turned out to be more extensive and expensive than predicted. Initial plans called for the Dock Street Theatre to open on February 12, 1936, to mark the institution's two-hundredth anniversary. That month, however, Mayor Burnet Maybank wrote Ellington that he believed November or December would be the best time to open the theater. "There is no possible chance of them finishing the job here before summer or very late spring," Maybank wrote, "and with our climatic condition it would be dreadful if we attempted to open the theatre at this time as we would be without our northern visitors and without our tourist business."[89] The mayor's words indicate the central role the city hoped the Dock Street Theatre would play in the tourism industry once opened. Indeed, local organizations also realized the restoration project had become an attraction to visitors, both out-of-town and local alike. The Society of Colonial Wars in the State of South Carolina erected a sign at the corner of Church and Queens Streets, with architect Ellington's permission, to provide information to the "hundreds of people" who stopped to look at the buildings every day.[90]

During the summer of 1936, architect Samuel Lapham reported that construction was "moving more rapidly." The roof work had been contracted, the concrete for the promenade around the rear and sides of the auditorium poured and the floor graded, and various steel and wood forms put in place for further construction work.[91] The timeframe of the project was soon extended, however, most likely as a result of construction challenges posed by the meticulous and careful adaptive use of the Planters' Hotel, the labor turnover common to New Deal relief projects, and changes in work relief administration as the government phased out the FERA and established the WPA.[92] As efforts continued over the winter and into the next spring, the WPA district director thought the project could still be completed by the end of June 1936, but architect Albert Simons told the WPA state administrator that six months more were required before the project could be "wound up." He admitted, "We have been very much concerned over the slow progress of this job . . . owing to the large scope of this work and its very complex and varied nature."[93] At times, Simons complained of the lack of appropriately skilled laborers, attributing this trouble to the "considerable volume of private construction now going forward in this locality." He urged Maybank to write to his friend Harry Hopkins and make the Dock Street Theatre work on a parity with private work: "This I should think might be based on the grounds that this job has already justified itself in furnishing relief work through the depression and now that we have passed out of this period of distress there exists a local emergency requiring the prompt completion of the building for immediate use. What I most fear is that if the completion of this theatre is indefinitely postponed at a time when it is greatly needed, public opinion will react most unfavorably towards the whole project and those of us who have sponsored it."[94] In response to this labor issue and potential negative publicity, Hopkins suggested Maybank "take skilled workers off other WPA jobs in Charleston," like plasterers and carpenters from the Navy Yard project, to get the theater project finished.[95]

Throughout the two-and-a-half-year restoration, Maybank requested additional funds from the WPA to supplement its original allocation for the expensive project. The architects were conscious of the high price of salvaging historic material, rehabilitating an existing structure, and constructing a modern theater on top of paying relief workers; indeed, Simons told Ellington, "We were anxious in every way possible to limit the cost of the work, as we were afraid that the funds might be exhausted before the entire project might be completed."[96] Yet, they need not have feared running out of money, for the mayor and architects' multiple requests for additional support throughout the project's duration from the WPA state administrator and chief engineer successfully resulted in increased federal grants.[97] By the time the Dock Street Theatre was ready to open its doors in November 1937, the federal government had allocated an astounding $350,000 to restore the historic site, perhaps the single most expensive WPA shrine restoration.[98]

In summary, Simons reported, engineering preparations necessitated over 25 sheets of architectural drawings, more than 100 sheets of architectural detail, and around 25 sheets of structural, electrical, mechanical, and heating plans.[99] The building also required 64 tons of structural steel, an acre of flooring, 8 miles of wood strips for plaster, 530 tons of concrete, and "uncounted bricks, kegs of nails and gallons of paint." Over the course of the project, workers removed more than 1,500 truckloads of rubbish from the site and donated 100 truckloads of flooring and framing to be used as firewood.[100] A WPA publicity release sent to hundreds of daily newspapers in November announced the project's completion and highlighted the engineering problems that arose during the construction phase. The WPA undoubtedly capitalized on the technical feats faced and overcome by the architects so that it could propagandize the project to which it contributed an extraordinary amount of federal funds. For example, to reinforce the building without removing the existing walls, workers developed a "special technique" that required digging six-to-eight-feet-deep pockets beside the solid standing walls, which allowed them to remain intact.[101] In emphasizing the construction challenges, the architects and WPA officials positioned the theater as a national example of masterfully executed architectural restoration.

Not everyone, however, was enamored with the theater's reappearance. Criticism of the project reveals that contemporaries understood the creative liberties taken by the architects and some questioned the value of investing enormous funds into inventing historic spaces. One such critic was journalist R. P. Harris, a preeminent editorial writer based in Baltimore. In an article for the *Baltimore Evening Sun*, Harris wrote, "Surely, the architects had a delightful time with this project. They ransacked decayed mansions for lovely doors and panels, they experimented with plaster and paint and ornamentation, to achieve what seems beyond question the right effect." Harris's language makes the salvage effort praised by Charlestonians sound like a plundering of the city's treasures. He also predicted that in a city of less than 65,000 people and with most out-of-towners preferring to visit plantations and beaches, there would not be enough interest in theatergoing to justify the cost of construction or sustain an audience once open.[102] In the midst of the Depression, Harris's concerns about spending capital on extravagant entertainment endeavors were understandable and shared by some locals who believed federal money would be better invested in other kinds of public works projects. The *Charleston News and Courier*, the city's leading conservative newspaper, published a few critical views of the theater restoration. In the mid-1930s, the paper was under the editorship of William W. Ball, a politically conservative Democrat notorious for his vocal and ardent opposition to the New Deal.[103] Ball's paper printed the opinion of one resident, who after an inspection of the Charleston County jail criticized money being spent on an "unnecessary project" like the Dock Street Theatre when funds could be used to improve prison conditions.[104]

Moreover, support for federal funding of the arts was not universal. Another resident called it "no business of government in any circumstance to hire unemployed musicians, artists, actors, [and] writers."[105] This comment speaks to the larger wave of criticism leveled in the mid- to late 1930s against the WPA's Federal One programs, particularly the politically controversial Federal Theatre Project, which led to congressional scrutiny and the program's early end in 1939.[106] As the Dock Street Theatre was sometimes mistaken for an FTP project, it makes sense that ire about federal dollars spent on the arts would be directed at the theater. Even some residents supportive of work relief art projects expressed disappointment; a member of the International Alliance of Theatrical Stage Employees and Moving Picture Machine Operators was chagrinned by the WPA's hiring practices, complaining to the relief district director and FTP director Hallie Flanagan, despite her having no involvement with the theater, that unemployed members of his organization should have been hired to complete the Dock Street Theatre's stage work before nonunionized members.[107] Despite these voices of discontent, the Dock Street Theatre generally received effusive admiration and praise from Charlestonians, government officials, and the press. Holger Cahill, national director of the WPA's Federal Art Project, even named the theater one of the most important federal projects in the state.[108]

As the theater neared completion, the city of Charleston considered options for its permanent management. The Ways and Means Committee of the city council recommended that the Carolina Art Association (CAA), which had managed the popular local Gibbes Art Gallery since 1905 and whose members boasted deep roots in the city, take over management of the theater following its opening on the condition that the association raise $12,000 from private individuals for operation and maintenance costs. Just over two weeks before opening night, the CAA entered a two-year contract with the city to manage the Dock Street Theatre for one dollar a year.[109] Upon assuming control, CAA president Robert N. S. Whitelaw said it was his desire to make the theater "a vital part of the life of the community and in no sense a stagnant 'museum piece,'" echoing Simons's words a year and a half prior that the theater not become "a museum piece, exquisite, but useless."[110] The CAA imagined that local art organizations such as the Footlight Players, the Poetry Society of South Carolina, the Musical Art Club, and the Society for the Preservation of Spirituals would be eager to use the theater in the years to come. The CAA, like Maybank, expected the expensive new Dock Street Theatre to spark interest in and then help sustain Charleston's abundant cultural offerings, which it hoped would find appeal both locally and beyond the city's borders.

Opening Night Welcomes Charleston's Blue Bloods

While the city of Charleston and the CAA negotiated the theater's management, the Footlight Players, an all-white amateur acting troupe, began readying for the first performance of George Farquhar's *The Recruiting Officer*. During the last two years of construction, the Footlight Players had loaned its warehouse to build the scenery and store equipment rent-free to the WPA project.[111] The South Carolina FAP oversaw set production. Although many FAP workers helped create the sets, two contributors of significance were Edith I. Allen, a junior architect of the Architectural Research Section of the Division of Women's and Professional Projects, and local artist and actress Alicia Rhett, who directed the painting of scenery.[112] The stage's main backdrop was based on prominent nineteenth-century artist and playwright John Blake White's 1838 painting of Charleston, which depicted Broad Street's St. Michael's Church, the old U.S. Custom House, and other notable landmarks (see fig. 4).[113] While working on the Dock Street Theatre's props, Rhett successfully auditioned for the role of India Wilkes in the film adaption of *Gone with the Wind*, the famously romanticized acount of the Civil War South, drawing further attention to the theater's restoration.[114] Meanwhile, women in the nearby WPA sewing room on Queen Street spent many weeks making the stage and balcony curtains, whose rust-colored velour with dull gold edging "blend[ed] beautifully with the Cypress paneling" of the balcony boxes and "add[ed] greatly to the old fashioned appearance."[115] This decor adorning the creatively engineered interior added important finishing touches to the bewitching portrait of a past that reinforced the same mythology of the South as *Gone with the Wind*.

The dedication to perfecting the illusion of Old Charleston, however, required specialized labor beyond the abilities of WPA relief workers. The architects turned to local talent to execute some design plans, hiring two skilled artisans to lend their expertise. The first was seventy-two-year-old John Smith, "a negro artisan not on relief rolls." Smith had previously restored the plasterwork on St. Philip's Church in 1920, and was hired to decorate the theater's elaborate plaster cornices and ceiling.[116] According to the superintendent of construction, Smith was "the only man in Charleston who knows how to do such work," though he was not given credit in any official agency literature about the project.[117] The second skilled artisan employed was William Melton Halsey, a young native Charlestonian who had studied fresco painting under renowned artist Lewis Rubenstein while attending the prestigious School of the Museum of Fine Arts in Boston. Simons gave Halsey his first commission as a professional artist by hiring him to work on the Dock Street Theatre. According to Halsey, Simons "was very concerned that everything should be in period so that none of what I did was original."[118] Consequently, Halsey decorated the courtyard fresco with a design based off of the bas relief of the proscenium arch in the old Academy of Music in Charleston (fig. 6), and adapted paintings by eighteenth-century English artist William Hogarth to complete the four oil murals in the barroom.[119]

The exterior preserved, the interior gutted and rebuilt, the construction challenges overcome, and the decorations set, all was prepared for opening week. After thirty months of construction, from May 1935 to November 1937, the Dock Street Theatre restoration finally came to an end. The CAA issued one thousand invitations to the nation's leading political, military, educational, and cultural figures for the first two invitation-only performances of *The Recruiting Officer*, scheduled for Friday, November 26, and Saturday, November 27.[120] Taking their esteemed seats on Friday evening were federal representatives Harry Hopkins, WPA director; Ellen S. Woodward, then

FIGURE 6: The Dock Street Theatre courtyard with a view of Charleston artist William Melton Halsey's fresco. Source: Frances Benjamin Johnston, Carnegie Survey of the Architecture of the South, c. 1936–37, Library of Congress Prints and Photographs Division.

WPA assistant director of the Division of Women's and Professional Projects; and Nikolai Sokoloff, director of the Federal Music Project.[121] No invitations were sent to Charleston's Black residents since the theater was a segregated space of the Jim Crow South. Leading up to the theater's unveiling, the WPA's central office drafted an announcement to be released to newspapers nationwide: "Hopkins to Dedicate Historic Theater Restored by WPA Workers."[122] Excitement brewed in major cities across the country as Charleston prepared to reveal its reconstructed colonial gem. Editors from the Associated Press, *Washington Post*, and *Time* magazine sent letters to Albert Simons and the CAA requesting information and photographs of the theater. Closer to home, the *State* newspaper in Columbia congratulated Charleston on its success, which represented "the springing of a memorable past into this present New Deal modernity."[123]

The opening evening's scheduled pomp and circumstance dazzled the special invitees, as was intended. Organizers created an ambience as grand and sparkling as the new theater in which they were to be entertained. Dressed in period garb, Citadel cadets and College of Charleston students served as ushers and escorted guests to their seats. They presented attendees with three souvenirs celebrating the theater's completion after two and a half years of research and construction: a commemorative booklet prepared by Douglas Ellington, the program to *The Recruiting Officer*, and handheld fans for the women. Roberta Maybank, the mayor's daughter, distributed fans similar to those ladies would have used in 1736 to conceal their blushes during the risqué performance.[124] These period-appropriate effects turned audience members into performers themselves as they adopted the behavior of colonial gentility. Once guests found their seats, the evening formally began with a Charleston String Symphony concert illuminated by candlelight.

Following the musical performance, WPA director Harry Hopkins presented the key to the theater to Mayor Maybank, enacting the giving and receiving of the Dock Street Theatre as a "gift" from the federal government to Charleston, a gesture affirming the WPA's substantial investment—both financially and ideologically—in the preservation of the local landmark.[125] In his speech, Hopkins related to the audience his affection for the southern city and acknowledged its distinct heritage:

> There is no city in America where this could have been done other than Charleston. This city has escaped the ruthless march of the

industrial system. Here a heritage of culture and arts is honored and respected. In dedicating this theater I would dedicate it to the people of Charleston—proud, fearless, courageous, intelligent. You have accepted faithfully a proud heritage and I believe your children and your children's children will accept the same heritage from you untarnished. Two hundred years from now our descendants may sit in this very theater. . . . It gives me great pleasure to present to the mayor of this city the key to this theater on behalf of the United States government.[126]

Hopkins certainly romanticized the city's illustrious past and unique claim to resisting modernity; Charleston had *not* escaped the march of industrialization. Nevertheless, he accurately voiced the city's deep commitment to wielding its built heritage as a tool to educate and entertain future generations. The Dock Street Theatre was significant not only to Charleston's history but to that of the United States as signified by Hopkins's use of the phrase "*our descendants*" and the suggestion that Americans across the nation would find value in the historical space.

After Hopkins's speech, the play's lead actor recited a prologue written by DuBose Heyward specifically for the evening before the Footlight Players launched into the performance of *The Recruiting Officer*. The Society for the Preservation of Spirituals, which by the 1930s had gained national attention through tours and radio appearances, closed the evening's program with ten of its most famous songs. In the words of one audience member, "the concert provided an appropriate climax to a night filled with exact reproduction of life in Charleston two hundred years ago," indicating that the showmanship was successful in turning the theater into a spatial memory cue.[127] Following the spirituals, Maybank invited the audience to inspect the theater, which led to "marveling at the remarkable craftsmanship shown in the restoration of the old building, its beautiful mantels, paneling, woodwork and architectural strength and beauty."[128] This tour of appreciation highlighted the role of the relocated Radcliffe-King House pieces in maintaining the illusion and marked a fitting end to the evening that reestablished Charleston's significant theatrical tradition. Before leaving the city, Hopkins told a reporter that the Dock Street Theatre was "one of the great institutions of America."[129]

The press coverage of opening night was overwhelmingly favorable, and news of the restoration spread far beyond the South. On the Sunday following the second performance, the *New York Times* published a column and two pictures of the theater. The following month, *Life* included seven pictures from the premiere, captioning them "First U.S. Theatre Is Restored: Charleston Blue Bloods Give It Gala Opening." The first *Architectural Record* issue of the next year contained twenty-two illustrations of the restored theater, and that same month Charleston schoolteacher Daisy Mae Roberts's article lauding the project for representing "a perfect blending of the atmosphere of the past with the ingenuity of the present" appeared in *Scholastic*.[130] Hopkins, Maybank, Ellington, Simons, and Lapham could find proof in these editorials that they achieved their goal of recreating a powerful visual of Old Charleston. At the same time, their work addressed a contemporary audience and sowed seeds of anticipation that Charleston would emerge from the Depression—with help from the federal government—as a cultural capital of the South.

A month before the theater opened to the public, Edmund P. Grice, the director of the WPA district office, remarked that "the whole idea behind this project was that it should be a contribution to the cultural development of the community. It is in no sense a commercial venture."[131] Aided by a Rockefeller Foundation grant of $15,000, in its first three years of operation the Dock Street Theatre developed its technical direction, welcomed touring companies from Europe, hosted a foreign film series, and opened a theatrical arts school.[132] As hoped for when plans for the restoration first materialized, various local art and cultural organizations made use of the performing arts space, including the Charleston String Symphony, the Charleston Philharmonic, the Poetry Society of South Carolina, the Dramatic Society of the College of Charleston, the Junior League of Charleston, and the Society for the Preservation of Old Dwellings.[133] Despite the recommendations of theater experts to employ professional actors and staff, as well as the continued and flattering interest of FTP director Hallie Flanagan, theater operations remained community-based with the local Footlight Players retaining its position as resident troupe.[134] Moreover, Charleston's favorite literary figure DuBose Heyward lent his fame to the institution by assuming the position of writer-in-residence from 1937 until his death in 1940.[135]

The city's cultural elite evidently succeeded in instilling a reverence for the past in visitors by attending to the preservation of pre–Civil War historic structures. When renowned writer of horror fiction H. P. Lovecraft visited Charleston in January 1936, in the midst of the Dock Street Theatre project, he expressed his appreciation for the city's architecture in a letter to a friend.

In his written "systematic itinerary" of a walking tour of the downtown area, he ignored the dire economic troubles the city faced and instead admired the "unique colonial features" of the grand homes and buildings actively maintained. Lovecraft venerated famous local sites, marveling at the ongoing transformation of the old Planters' Hotel into the Dock Street Theatre and the Georgian buildings of Vanderhorst Row on East Bay Street, home to some of the city's oldest and wealthiest families. Because of this "encouraging reclamation" of historic landmarks funded by New Deal programs, which Lovecraft hoped "may not be interrupted by any reactionary political move," Charleston remained "refreshingly free from 'modernistic' architecture." After taking in its charm and history, the writer remarked that the city "is alive in every sense despite the omnipresent aura of the past."[136]

In the souvenir booklet published to commemorate the Dock Street Theatre's grand opening, the director of the South Carolina FAP reminded contemporary audiences of the restoration's pragmatic appeal and idealized charm: "The story of this reconstruction is a chapter in the greater story of the government's program of work relief. But it is also a chapter of compelling romance."[137] By shaping Charleston's historical image through fancifully engineering the reappearance of the eighteenth-century Dock Street Theatre, Charleston's political and social elites pushed forward an agenda to safeguard structures that reinforced racial and class hierarchies. However, while the theater emerged as a segregated space in the Jim Crow South, the project was not effective in entirely removing Black Charlestonians from the area. A close look at the center of figure 3 reveals that when the photograph was captured in 1936 or 1937, both white and Black residents still moved through the public space of the old Planters' Hotel's historic corner of downtown. Standing on opposite sides of Church Street, a Black woman pushes a baby carriage while a young Black male standing next to a white figure looks on. Perhaps the woman is a hired nurse, but these two figures' presence outside of the theater, then still under restoration, belies the myth of a historic white Charleston perpetuated through the theater project.

Nonetheless, the New Deal preservation project *did* succeed, inarguably, in both ideological and practical ways; it enacted Mayor Burnet Maybank's liberal plan to modernize his city by capitalizing on newly available funds for civic infrastructure, and it helped a group of Charlestonians intent on maintaining control over cultural heritage reinvent the architectural character of the southern city's historic core. The Dock Street Theatre that reemerged in the 1930s, then, importantly expressed an entrenched conservativism and simultaneously defended preservation as an integral part of community revitalization. While the theater's construction advanced southern Democrats' maneuverings to harness New Deal resources, it also allowed Charleston's white elites to tangibly craft a palatable representation of the city as a nostalgic bastion of their past privileged lives. The political and art leaders responsible at the local level successfully redefined and cultivated a cultural identity of Charleston promoted prominently in white tourism in the twentieth century and beyond.

CHAPTER 3

The Puritan Past at the Old Stone House

The Henry Whitfield State Museum, Guilford, Connecticut

When one looks today at the fine old house, restored as nearly as possible to its original condition, firmly planted in the earth where it first took root, we are grateful to have emerged with even some measure of success. We are happy to have been one of that long line of lovers of this old house, interested to save it from destruction or decay, and our hope is that those who may come to study and enjoy it and its surroundings will find that it measures up to the best scientific standards—archaeological, historical, architectural—known and used by architects and scholars of today.

Evangeline Walker Andrews, chair of the Restoration Committee,
Henry Whitfield State Museum, October 20, 1937

In the fall of 1937, the Henry Whitfield State Museum in Guilford, Connecticut, opened its doors to an eager public after sixteen months of extensive renovation (fig. 7). During that period, renowned restoration architect J. (John) Frederick Kelly gutted and then rebuilt the "Old Stone House" of 1640, which had served as the state historical museum since 1900. This project, directed by the museum's Board of Trustees and funded by the Works Progress Administration, returned the town's—and possibly the state's—oldest stone abode to its seventeenth-century appearance when its first and most prominent resident, the Puritan Reverend Henry Whitfield, lived there. Formerly minister

FIGURE 7: Front exterior view from the west of the Henry Whitfield State Museum, 1937. Source: Historic New England.

in the village of Ockley in Surrey County, Whitfield and twenty-four others embarked on a voyage from England to the Connecticut Colony in the spring of 1639. Fleeing King Charles I's persecution of Puritan clergy, the party first landed at New Haven. The English settlers, however, chose to establish their own town halfway between New Haven and Saybrook, and purchased land from the native Menunkatuck peoples, which they renamed Guilford. After the settlers' arrival in late fall, they began building the first permanent structure in the new town, Whitfield's home, a few blocks south of what would become Guilford town green and about one mile north of Guilford Harbor on Long Island Sound. Most likely completed in the spring of 1640, Whitfield's house allegedly functioned as the town meeting hall, a place to shelter from Native American attacks, and the community's religious center until the colonists built the nearby Congregational Church in 1643.

Whitfield's residence was a sturdy stone house, similar to the kind the settlers had known in their native English counties of Surrey and Kent. Legend has it that the colonists transported heavy stones from a nearby quarry in handbarrows with the help of the native Menunkatucks, perhaps not realizing that plenty of wood was available a mere quarter of a mile away or choosing stone over wood because it is a stronger material. In construction, the builders used pulverized oyster shells and yellow clay for mortar, hand-hewn oak timbers for the beams, and wide planks of native pine for the floors.[1] The family's house on its nine-acre plot was one of only four stone houses built by the earliest settlers in Guilford and the only one still standing by the twentieth century.[2] It is possible that its formation, which probably included a surrounding stone wall, was modeled on the "bawn," an enclosed defensive structure the English built in their Irish colonies during the early seventeenth century.[3]

The finished home's first occupants included Whitfield; his wife, Dorothy; and some combination of their nine children and a few servants. In 1651, the Whitfields returned to England, and for the next almost 250 years the Old Stone House remained a private residence lived in mostly by tenants rather than its owners.[4] Then, near the turn of the twentieth century, the property changed from private to public ownership when the Connecticut Society of the National Society of the Colonial Dames of America (Connecticut Society) successfully campaigned for the state legislature to acquire the property and convert it into the state's historical museum in 1900, making the Whitfield House one of the earliest house museums in New England and likely the first owned by a state government.[5]

During its first couple of decades as the state museum, the Whitfield House withstood both physical alterations when the structure was converted to accommodate exhibition space and cultural shifts as first the "Roaring 1920s" and then the Depression transformed economic and social conditions in Connecticut. Significantly, the change in function of the Old Stone House from private residence to public museum accompanied an overall decline in the social capital extended to New Englanders simply because of their association with Puritan forebearers. Throughout the first few decades of the twentieth century, intellectuals in the South and the middle states increasingly resented New England's prominence in narratives of national history, which typically began with permanent English settlement on the Atlantic Coast, and the region's enduring cultural dominance. New Englanders themselves struggled to extol the stifling standards and humdrum of the Puritan model amid the gaiety and seeming prosperity of the 1920s.[6] Indeed, historian Michael Kammen argued that "Puritan-bashing" became a popular pastime in the 1920s, and Karal Ann Marling explained that criticizing the Puritans "amounted to a mark of cultural maturity."[7]

The alarming crash of the stock market in 1929, however, engendered a reevaluation of the decadence and debauchery of the previous decade and a reconsideration of the Puritan stock character's usability. The title of "Puritan" and the historical imagery it called to mind came back into fashion during the more reserved years of the Depression. Literary critic Van Wyck Brooks, once responsible for the "retrospective debunking" of the Puritans, reconsidered the utility of their stern, moral ways and published the bestselling, five-volume "Makers and Finders" series, beginning with *The Flowering of New England, 1815–1865*, in 1936.[8] According to cultural historian Warren Susman, the idea of the Puritan past centered around four issues that made it particularly "usable," or relevant, in the 1930s: God-centered self-restraint and control, a dedication to community and law and order in a rigid social system, a strong sense of morality and strict code of ethics, and material success begotten through thrift and industrious efforts.[9] The characteristic austerity of Puritan society matched the soberness of the Depression era, and the Henry Whitfield State Museum became perfectly poised as an educational institution to instruct visitors in the renewed veneration of the town's ancestors.

Changes in Connecticut state politics reflected the variable cultural and economic climate of the period. In the 1920s, along with other northeastern and mid-Atlantic states, Connecticut witnessed the rise of an urban-industrial society. Increased numbers of immigrants from eastern and southern Europe, Jews, and Catholics fundamentally altered state demographics and threatened the supremacy of the Protestant Yankee, an outgrowth of the original Puritan character. Meanwhile, "Yankee Republicans" ran state government like an efficient machine, encouraging the growth of business and consumerism.[10] The beginning of the Depression and growth of the Democratic Party in state affairs during the Roosevelt years represented a break from the old political ways of the Republican bosses. Governor Wilbur L. Cross, former dean of the Graduate School of Yale University, entered office in 1930 as a Democrat with a reform platform, challenging, but not completely overturning, the long-held Republican rule of the state. Serving from 1931 to 1939, Cross remained a "circumspectly liberal Democrat": naturally wary of federal spending, a proponent of states' rights, and prudently suspicious of the New Deal's extension into state and local politics. While for most of his time in office he faced a Republican state legislature, he successfully led Connecticut away from "the indolence of the Republican machine" by enacting popular reforms. Major projects included a road-building program, flood relief measures, and reform legislation to abolish child labor, institute a minimum wage, and increase public utilities.[11] With Cross at the helm, Connecticut's Democratic Party transitioned from the "Old Guard" urban bosses to the "New Guard" and finally secured control of the state legislature in 1936, the same year the state voted to reelect President Roosevelt to his second term.[12]

Historian Robert L. Woodbury described Cross's savvy steering of the Democratic Party in the 1930s as a push to bring "a 'little' New Deal to a state with a deeply rooted Yankee heritage."[13] Indeed, part of Cross's appeal was his respectable Yankee background, meaning his illustrious Protestant New England ancestry. The governor belonged to a family that could trace its lineage back three centuries, grew up in the small rural town of Gurleyville, and studied at Yale University, where he nurtured his belief in ideals of anti-imperialism, anti-Catholicism, and laissez-faire economics.[14] His background helped soften the Democrats' image at the state level, and his ideologically conservative pledge to a balanced budget mollified some Republicans.[15] In his upbringing and politics, then, Cross embodied the conservative values and work ethic of the Puritan Yankee stock figure; he espoused individual responsibility, free will, and strong moral fiber. A friendly and skilled politician, he "conceived of his own role not primarily as that of a political leader, but as that of a nonpartisan moral and spiritual mentor of Connecticut," Woodbury argued. While Cross challenged Republican political power, he did not represent a rupture from the state's cultural and historical traditions. His public speeches, moreover, revealed "a self-conscious debate between the historical conservative and the uncomfortable liberal, an interior dialogue between the values of his heritage and the casualties of a depressed urban and industrial society."[16] Thus, while a modern liberal in some regards, Cross equally identified as an exemplar of a deep-rooted, historically conditioned "Connecticut Yankee," a title he willingly embraced and gave to his published autobiography in 1943.

The political dichotomy between conservative, heritage-based values and Governor Cross's mild, New Deal liberal agenda manifested in the WPA's restoration of the Henry Whitfield State Historical Museum in Guilford. While the work relief project clearly represented the extension of the federal government into local affairs, a symbol of New Deal federalism, the restoration also resulted from a powerful and popular nostalgia for the much-mythicized Puritan past and the qualities long associated with the early English settlers: self-restraint, morality, and hard-earned success. It thus represented a rejection of the ongoing and permanent shift toward modernization, urbanization, expanded federal power, and a more diverse and democratic citizenry—the hallmark forces of the New Deal era. This desire to reinforce the town's earliest chapter of Anglo history also firmly placed the Whitfield House within the context of the prevalent colonial revival movement and regional "Yankee progressivism." The local urge to use architecture to promote a Puritan mythology exemplified the tenets of traditionalism—belief in local government, laissez-faire capitalism, private property, individualism, family values, patriotism, and work ethic—that devolved from cherishing imagined colonial landscapes, as James M. Lindgren argued.[17] During the first few decades of the twentieth century, both professional and amateur admirers of historic homes grew to regard the architecture of early chapters of American national history as a critical educative force in "Americanizing" foreigners. Women's groups like the Colonial Dames of America and professionalizing groups like the regional Society for the Preservation of New England Antiquities (SPNEA) preserved seventeenth- and eighteenth-century homes, churches, and government buildings and created quintessential colonial villages that appealed to residents and tourists alike, such as the towns of Litchfield and Guilford in Connecticut.[18]

The colonial revival impulse guided two separate restorations of the Whitfield House by leading professional New England architects: the first undertaken by Norman M. Isham and funded by the state of Connecticut between 1902 and 1904, and the second completed by J. Frederick Kelly as a WPA work relief project from 1935 to 1937. Kelly's work was motivated by changes within the preservation field that prioritized modern management of projects, scientific study of architecture, and masculinized professionalism. He thus employed a scientific approach to restoring the Old Stone House, relying on physical evidence and architectural documentation to inform his decisions rather than feelings of associationalism, a move that reflected a significant trend in preservation practice exemplified by the institutional work of SPNEA and the Colonial Williamsburg endeavor in Virginia taking place contemporaneously.[19] But Kelly's professional choices highlighted the tensions that resulted from the transition from amateur to professional, and relatedly women to men, within the preservation movement. When WPA laborers, under Kelly's leadership, began restoring the Old Stone House, the town's most sacred structure, townsfolk were initially apprehensive, and some became openly hostile. Although Kelly was a reputable colonial architecture expert and executed a well-researched, technically sound restoration, women who were involved in the local Dorothy Whitfield Historical Society and the Connecticut Society eyed with suspicion the changes he made that altered the house's appearance. Kelly and the museum trustees' pursuit of historical "authenticity" at the Old Stone House conflicted with town residents' memories of the site, and a battle was waged between the architect and museum trustees and the town's amateur memory keepers, who long had served as protectors of its historic homes.

Besides exemplifying shifting power relations within the preservation field, the Whitfield House restoration reflected an ongoing struggle to administer a federal public works project in a town that distrusted big government. Kelly and the trustees, many of whom were members of historical and preservation organizations, were used to managing projects in the private sector and found it difficult to adhere to the WPA's regulations regarding hiring practices, acquisition of materials, and work schedules. Beyond that, Guilford residents negatively associated Kelly's professional decisions with the WPA as his employer, finding reason to be wary of both federal intervention and outsider-expert influence in the restoration of a site around which Guildford had built its historical identity for generations. The challenges the architect and trustees encountered while navigating the agency's ever-changing rules thus were confounded by local fear that the intrusion of big government would result in increased state control over the use and public interpretation of a locally revered site.[20]

Despite simmering tensions, both town residents and the project leaders emphatically agreed on the restoration's goal to reinforce the importance of the town's Puritan history through architecture, a renewed symbol of survival, persistence, and conservative commitment in a region that tepidly embraced the politics of the New Deal. They cared deeply about the Whitfield House and the message it sent as the state museum to both locals and out-of-town visitors because they believed the historic building provided narrative stability as well as tangible roots to their community. They held that it was the proprietary right and responsibility of concerned locals, rather than distant government figures, to maintain the historical legacy of the Old Stone House. Exploring the rocky unfolding of the WPA project to restore the Henry Whitfield State Museum and the tense communication among project leaders, professionals, and residents illustrates the continued power of Puritan mythology as a nostalgic defense, the friction that accompanied professionalization of the preservation field, and resistance to extended state control into local affairs.

The Old Stone House Becomes the State Historical Museum

By the time the WPA considered the Whitfield House a "historic shrine" in the 1930s, New Englanders long had celebrated and commemorated the site for its historical association with the region's Puritan past. Throughout the eighteenth and nineteenth centuries, evocations of the Old Stone House sheltering colonists during Native American raids, bringing settlers together in prayer, and bearing witness to Guilford's earliest ceremonies and festivities as the informal town meeting hall endured. The *North American Tourist*, the nation's first comprehensive guidebook, encouraged travelers to visit the site (1839); the popular *Ladies' Repository* periodical published a steel engraving of the home (1863); and *Potter's American Monthly* featured an illustration and recounted the tale of stone being carted in handbarrows to build the two-feet-thick walls (1875).[21] Townsfolk sold German-made plates with views of the Old Stone House, wooden crosses made from old beams, and postcards as souvenirs to promote Guilford's local landmark as a revered symbol and popular tourist attraction.

In the last quarter of the nineteenth century, the romanticization of the Whitfield House only escalated with the growth of the colonial revival movement. At the Centennial Exposition of 1876 in Philadelphia, visitors moved through the fabricated historical Connecticut Building and New England Log House. Later, even larger numbers of people toured the state pavilions and exhibitions at Chicago's Columbian Exposition in 1893. The world's fair recreated Massachusetts's John Hancock House, Virginia's Mount Vernon, and Connecticut's simple "farmhouse," which though humbler than the other designs on display was still grander than its historical prototype. Both the general public's and the architectural community's interest in the colonial period blossomed, and prominent architects Richard Michell Upjohn, Robert Swain Peabody, Charles Follen McKim, and others began to design domestic and commercial buildings in colonial styles and publish writings on colonial architecture.[22]

The colonial revival movement developed to some degree as a reactionary response to the growing ethnic, religious, and linguistic diversity resulting from increased immigration, especially from eastern and southern Europe, in the period preceding World War I. Typically, rather than being explicitly xenophobic, proponents of the colonial revival employed architecture and material culture to teach and mold immigrants into good American citizens, with the colonial home and hearth at the forefront of Americanizing efforts.[23] Colonial styles recalled the refined manner of living during the nation's period of western European settlement, the morals and manners of which became a compass for newly arrived immigrants to follow. Major anniversaries, like the 1920 tercentenary of the Pilgrims' landing in Plymouth, Massachusetts, provided commemorators the opportunity to "harness the power of heritage" to bolster America's Anglo-Saxon Protestant identity and address what they considered challenges to democracy, namely immigration and socialism. Government-funded improvements to Plymouth's historic area and a grand historical pageant helped turn the landing site into a "shrine" for modern American pilgrims.[24]

As explained in chapter 1, prominent patriotic and hereditary organizations whose members could trace their ancestry to the colonial period, like the National Society of the Colonial Dames of America, popularized colonial images in historic homes, town celebrations, printed literature and memorabilia, and objects. According to Karal Ann Marling, hereditary societies like the Colonial Dames operated with a sense of "defensive nostalgia" for a past over which they were fiercely protective in the wake of perceived threats by newly arrived foreigners.[25] Founded in 1890, the Colonial Dames also actively preserved and converted old buildings into historic house museums that became part of the curriculum of patriotic education.[26] The Connecticut Society, formed in 1893, exemplified the kind of work the national organization performed regarding the management and care of historic properties in the name of patriotism and reassertion of English ethnocentrism when it transferred the Whitfield House from private to public ownership and established it as the State Historical Museum at the turn of the twentieth century.

Yet, while residents clearly held widespread affection for Guilford's Old Stone House, love and pride of the building itself did little to protect it from the expected wear and tear and financial burden that come with time and required upkeep of a historic house. By the 1890s, an out-of-state owner held several mortgages on the rundown property and demolition loomed as a threat. The Connecticut Society believed Whitfield's house to be the best representative of the state's founding Puritan ideals and was spurred to take action to save it. In 1897, purportedly encouraged by the lawyers representing the mortgages, society members launched a campaign to rescue the venerated home and turn it into the state museum, thereby ending over 250 years of use as a private residence. This act coincided with the preservation work of other Colonial Dames chapters; for example, in 1900 the Wake County Chapter of the North Carolina Society purchased Andrew Johnson's supposed birthplace for one hundred dollars, presented it to the city of Raleigh, and had it relocated to Pullen Park to be interpreted as part of the state's heritage.[27]

Two years after the Connecticut Society began campaigning, the state legislature passed an appropriations bill on June 22, 1899, approving the state's purchase of the Old Stone House. A year later, the widowed owner Sarah Brown Cone finalized the transaction when she sold the Whitfield House and eight acres of land to the state of Connecticut for $8,500. Upon receiving the title to the house and lands, the state agreed to appropriate $2,000 biennially for its administration and maintenance.[28] The state legislature also authorized the governor to appoint an eight-member Board of Trustees, which could accept external funds to be used for the upkeep and support of the house. According to this governing body, the state legislature decided to purchase the Whitfield property for use as a museum because it realized "its importance as a historical relic, unique in the fact that it stands alone as the original home of the leader of a colony and as the only stone house of its period in our country north of Florida."[29]

When the state acquired the Whitfield House, however, its appearance was a far cry from its original 1640 condition. Its form had undergone innumerable changes as subsequent owners enlarged and renovated the property. At some point in the eighteenth century, owners or tenants had removed the south chimney and converted the windows from casement to double-hung, widening them at the same time. Then, a fire sometime in the first few decades of the nineteenth century left the house roofless and unlivable for a period. The unprotected walls weakened to such an extent that when then-owners Mary and Henry Ward Chittenden began repairs in 1868, the south wall had to be entirely rebuilt and nearly half of the front (west) wall as well. When doing so, the Chittendens heightened the masonry walls, which flattened the new slate-covered roof and decreased its pitch. They also installed new floors, built a large ell at the rear of the house in the northeast corner, and renewed the exterior covering of stucco, which previous owners had applied in 1820 to help preserve the original masonry underneath.[30] In short, the Henry Whitfield State Historical Museum around 1900 hardly looked like the stone house the minister Whitfield called home in 1640 (compare figs. 7 and 8).

To restore the Whitfield House to its colonial-era appearance and make it suitable for museum purposes, the Connecticut Society's newly formed Historical Sites and Henry Whitfield Committee hired architect Norman M. Isham. He had established himself as an expert on seventeenth-century residential architecture with the publication of *Early Rhode Island Houses* in 1895, which he followed with *Early Connecticut Houses* in 1900. Tasked with transforming a private residence into a public museum, Isham did not intend to recreate the original interior as it would have appeared during Whitfield's residency. Indeed, he noted that "the house that Whitfield built can hardly be said to exist . . . save as a shell."[31] Consequently, he created what he called the "Great Hall" to present a higher style of architecture suitable for a museum gallery.[32] He converted the front rooms into a single, two-story exhibition

FIGURE 8: Henry Whitfield House, c. 1902, before architect Norman Isham's restoration. Source: Henry Whitfield House Restoration and Landscaping Projects, 1900–1940, RG024:001, State Archives, Connecticut State Library, Hartford.

space to display museum objects, installed a new floor of handsome oak, and adorned the ceiling with chamfered oak beams (fig. 9). The fourteen-foot-wide by thirty-three-foot-long hall had an imposing height of sixteen feet and featured folding partitions that could divide the room into multiple sections. The design of this new exhibit space opened the once-enclosed stair tower where Isham placed a newly constructed elaborate Jacobean staircase.[33] In the deep window openings, the architect installed new double-hung windows with diamond-shaped leaded glass panes that "add[ed] much to the look of antiquity," which he additionally decorated with carved oaken architraves.[34] Isham also restored the large original fireplace in the northern wall and designed a mirror fireplace at the southern end of the room that was purely ornamental for there was no stack or flue. He left alone the back section of the house, where lived caretakers Everett and Cornelia (or Amelia) Dudley, who had resided on the property as farm managers for owner Sarah Brown Cone since the 1870s.[35]

Isham completed the work in June 1904, and the State Historical Museum later held a formal opening on September 21 to welcome the public. A tablet commemorating the Connecticut Society's efforts to acquire the house for the state in 1897 was affixed to the facade. Free admission beckoned people to visit the museum to learn about the town's forebearers, interact with artifacts of yesteryear, and find inspiration in Puritan morals and behavior.[36] Over the course of the restoration, the museum collected more than two thousand items, either by donation or loan from other historical institutions, to display in the Great Hall, which became an elaborate period room. Historian Jessie Swigger argues that the "period room," popularized during the colonial revival, "became a unique vehicle for communicating patriotism, nostalgia, and nationalism."[37] At the museum, colonial-era furniture and items representing domestic skills like spinning wheels and looms, tableware, and framed artwork encouraged visitors to imagine and admire what life would have been like for their colonial ancestors (see fig. 9). Visitors could also move through some of the house's small "apartments," while the caretakers' quarters and some ancillary rooms were closed to the public. At the opening ceremony, Professor Samuel Hart, president of the Connecticut Historical Society, described the Whitfield House as "a place of historic witness" where visitors could learn about the original settlers "who laid in these colonies such abiding foundations."[38] The glorified foundations were both literal in the durable,

FIGURE 9: Norman Isham's "Great Hall," c. 1904. Source: WHI-(71499), Wisconsin Historical Society, Madison.

permanent stone walls that survived over two and a half centuries, and figurative in the Puritan template for moral living they enshrined.

Plans Unfold for the Re-Restoration of the Old Stone House

Norman Isham's restoration of the Old Stone House and its founding as a public museum marked a momentous achievement for the Connecticut Society. The Henry Whitfield State Museum was one of the earliest historic house museums in the nation, and it preceded by almost a decade the restoration of two other notable historic homes in New England that fostered "patriotism, Anglo-Saxonism, and acceptable Yankee values": the Paul Revere House in Boston (1908) and the House of Seven Gables in Salem, Massachusetts (1910).[39] Yet, despite the seemingly jubilant climate of the museum's opening, and the respect Isham's peers held for him in New England's elite preservation circle, criticism of the architect's restoration, which centered on his design decisions, began as soon as he completed the work.[40] When transforming the Old Stone House into an appropriate museum venue, the architect intentionally and honestly prioritized, in his own words, "a comfortable and dignified character as well as the *flavor* of the seventeenth century." A historically accurate reconstruction of the minister's abode, he professed, would have been impossible for lack of documentation. Thus, because of Isham's charge to create a historical exhibition space, he looked to grander mid-seventeenth-century English homes and medieval designs as inspiration over comparable modest American colonial structures.[41] The needs of the museum, he believed, surpassed guesswork at architectural accuracy in restoration work. This decision resulted in the creative reproduction of an imagined colonial space, much like the Dock Street Theatre discussed in the previous chapter.

George Dudley Seymour, a New Haven lawyer, hobby restorationist, and proponent of the colonial revival, quickly took issue with Isham's so-called restoration.[42] Shortly after his appointment to the state museum's Board of Trustees in October 1907, Seymour presented a formal paper in which he expressed his disapproval of Isham's work, calling the architect's gravest mistake ignoring the mid-nineteenth-century plans of the house produced by Guilford genealogist and historian Ralph D. Smith, published in John Gorham Palfrey's *History of New England* in 1859.[43] Isham had, in fact, consulted Smith's plans, but he chose instead to base his design on "tradition, inherent reasonableness, and likeness to old English examples."[44] Seymour also criticized Isham's two-story Great Hall, which he described as "sufficiently absurd" since it invalidated the need for a staircase to the second floor. He decried Isham's work as "far from supported by reliable traditional and evidential authority," and charged that as a result, the museum was "not a credit to the State nor to the Trustees."[45] Moreover, and injurious to the museum's educative goal, he claimed that "the house as restored is unconvincing to students of early [architectural] work and grievously disappointing to visitors," as it did not paint a truthful picture of colonial life.[46] As a devoted antiquarian, Seymour took umbrage with the museum's false depiction of the past, equating authentic architectural reproduction with the capability to accurately relate history.

Consequently, Seymour began in earnest a campaign for the "re-restoration" of the Whitfield House designed according to the 1859 plans drawn by Smith, whose grandson Dr. Walter R. Steiner recently had become a museum trustee. After more than two decades of campaigning, Seymour's persistence finally paid off when the Board of Trustees decided to move forward with a second major restoration in August 1930. The board hired for this task architect J. Frederick Kelly, who had converted an 1870s barn on the property into a caretaker's house back in 1923.[47] A graduate of the Yale School of Architecture, Kelly was a natural selection for the re-restoration. He ran a successful architectural firm in New Haven with his brother Henry Schraub Kelly, had begun a productive professional relationship with the New Haven Colony Historical Society, and had published several important writings on the architecture of early Connecticut, most notably *Early Domestic Architecture of Connecticut* in 1924.[48] Kelly's understanding of old buildings as instructive, unique, and deserving of veneration clearly aligned with the principles of Yankee progressivism upheld by the regional preservation movement. In an essay written for the celebration of Connecticut's tercentenary in 1933, the architect urged the preservation of the state's "ancient houses" as public policy and referred to them as "*human documents* of the greatest value and the utmost significance. They must not be destroyed, for they form a vital and irreplaceable link with a vanished past and a people whose part in the upbuilding of our nation merits our humble and reverent admiration."[49] Moreover, in calling buildings "documents" he articulated the professional opinion that restoration work required physical or textual evidentiary support. He viewed historic preservation as both an art and a science, prioritizing aesthetics and moving away from older iterations of preservation based purely on historical associationalism and patriotism.[50]

Kelly's relationship with two preeminent organizations within the fields of architecture and history bolstered his professional stature and shaped his views on historic preservation. The first was SPNEA, which formed in 1910 to preserve "for future generations the rapidly disappearing architectural monuments of New England and the smaller antiquities connected with its people."[51] Like the Colonial Dames, the organization purchased and maintained historic properties, some of which it rented out to tenants and others it operated as museums. SPNEA also published *Old-Time New England*, a scholarly journal devoted to the architecture, material culture, and general lifestyle of the colonial period to which Kelly frequently contributed as an author.[52] Sometime before 1918, Kelly began a professional relationship with William Sumner Appleton, SPNEA corresponding secretary and a powerful figure in the New England preservation community, and the two remained in communication throughout Kelly's tenure with the New Haven Colony Historical Society.

The second prestigious association to which Kelly belonged was the Walpole Society. Formed in 1910 like SPNEA, the Walpole Society similarly promoted the study of American history, architecture, and decorative arts largely through antiquities collecting and dealing. The very exclusive, all-white group was composed of leading male architects, historians, and collectors, and both professional and familial credentials determined membership.[53] In the 1930s, the society limited membership to twenty-five men, and each candidate had to be officially endorsed by two current members and unanimously voted in by secret ballot at a society meeting. Being of an old Episcopalian family from New York, Kelly possessed the desired professional and social qualifications. Importantly, he also had developed a productive working relationship with George Dudley Seymour, founding member of the Walpole Society. In addition to their work together on the Whitfield House, Seymour and Kelly collaborated on projects for SPNEA, the Gallery of Fine Arts at Yale University, and the New Haven Colony Historical Society.[54] Kelly's reputation as an outstanding architect, restorationist, and scholar was validated by his invitation, proposed by Seymour, to join the elite Walpole Society in 1935.

When Seymour first approached Kelly about undertaking a serious re-restoration of the Whitfield House, the architect expressed minimal interest. Back in January 1927, Kelly had explained to museum trustee Alfred Hammer that he "did not feel like attempting the work in a professional capacity," a position he reiterated when asked again a year and a half later. He reasoned that the house was unique as the only surviving mid-seventeenth-century stone house in Connecticut, and therefore he had no precedent to draw on for guidance. The initial resources he gathered on the house were "contradictory," and he knew it would be difficult to "sift out what is sound from the chaff." Moreover, Kelly predicted that "whoever does the work will be bound to be subjected to a great deal of criticism," probably referring to the treatment of his esteemed fellow architect Norman M. Isham.[55] Yet Kelly resisted continued requests to work on the Whitfield House on other grounds as well. He felt disinclined to involve himself with another property under the tutelage of the Colonial Dames, whose Connecticut Society by this time had been involved in the management of the house for three decades. Kelly had worked closely with the society in the past; he restored the Webb House in Wethersfield, which was owned by the Dames, and since 1915 had been commissioned by its Committee on Old Houses to make architectural drawings for many of its publications.[56] In a 1930 letter to museum trustee Hammer, Kelly explained, "Personally, I feel that it will be nigh impossible to carry on work of this nature under the supervision of a committee of women, and do it as it should be done. . . . By bringing into the picture a committee of no doubt well-intentioned women who know nothing about the problem in hand, seems out of the question, and I should not care to undertake it."[57]

Kelly's derisive characterization of the Connecticut Society as "well-intentioned" but ultimately lacking professional aptitude to continue directing preservation efforts at the Whitfield House substantiates a critical change in attitudes and practices within the national historic preservation movement that had occurred over the previous two decades. Educated professional men like Kelly, Seymour, and Appleton took the reins in architectural restoration as they introduced principles of business and modern management and marginalized women. Groups like the Colonial Dames, pushed out of their traditional roles in historical commemoration, shifted efforts to other heritage-based educational campaigns to further Americanism, relinquishing control of architectural restoration to the new so-called experts.[58] Regarding the Old Stone House in particular, Kelly's reluctance also may have stemmed from the fact that the Connecticut Society oversaw Isham's inaccurate restoration, and he feared members would influence his own plans if he were to take on the project since the museum was still "under the auspices" of the group. Regardless of the reasons for his hesitance, Kelly did not acknowledge that it was the Colonial Dames who first had protected the site by orchestrating the state's purchase of the property in 1899, thereby discrediting women's influential role.

Despite misgivings about the project, Kelly finally agreed to accept the commission to re-restore the Whitfield House in the late summer of 1930, probably as a result of the persistent urging and influence of his friend and colleague George Dudley Seymour. That same year, the Board of Trustees successfully petitioned the state legislature for a $10,000 grant, which allowed Kelly to begin implementing his designs for the property. His three-stage plan called for first reconstructing the original ell in the northeast corner, which Sarah Brown Cone's family had replaced with a larger ell, in accordance with the measurements of Ralph D. Smith's plans.[59] Kelly reintroduced two secret closets into the garret of the ell, which also appeared in Smith's drawings. The state grant additionally financed the second step in the restoration scheme, which took place in the summer of 1933 and involved removing the mid-nineteenth-century coat of stucco covering the masonry of the main part of the house. At this time, the architect wrote that existing seventeenth-century historic structures in and around Guilford guided his choices of the materials and "the general character of the work," which already distinguished his approach from that of Isham, who used more grandiose English models.[60]

With the completion of this second stage, funds ran out, and the Board of Trustees put together a proposal asking the state legislature for an additional appropriation of $20,000 to complete the re-restoration in preparation for Guilford's tercentenary celebration in 1939.[61] While the state legislature sat on the proposal, the board continued to develop plans for the third step of the work, which involved the most difficult task: changing the height of the walls and roof along with more interior alterations. On May 9, 1935, the board appointed a Restoration Committee, selected from the trustees and tasked with fundraising for this final and costliest stage.[62] As the new committee organized under trustee Evangeline Andrews, the board directed Kelly to prepare a booklet of his architectural drawings and detailed plans for the house to be published and sold at the museum in an attempt to garner attention and potential sources of funding for the forthcoming work.[63]

The WPA Enacts the Final Stage of Re-Restoration

Despite the Board of Trustees' and Kelly's efforts, the state legislature denied the requested $20,000. Dealt another blow, the board received only $2,500 from the State Tercentenary Commission a year later in June 1936.[64] While the board surely found the state's paltry contributions a disappointment, Evangeline Andrews, as head of the Restoration Committee, prudently had sought alternative funding sources before the state legislature even made its final decision. In the spring of 1935, Andrews began communicating with Eleanor Little, longtime resident of Guilford and relief administrator of the Connecticut Emergency Relief Commission (CERC), the state arm of the New Deal's Federal Emergency Relief Administration. Trustees Alfred Hammer and Frederick Norton already had brought the board's idea to restore the Whitfield House to Little's attention when they hoped it would be funded as part of the tercentenary celebration, but Andrews began to actively push the restoration as a CERC project while waiting to hear whether the state legislature would appropriate funds. Little quickly approved of the Whitfield House project application the following month and wrote Andrews that she was "hopeful that it can be approved without delay" by the FERA's Washington, DC, office.[65] While Kelly may have disregarded the efforts of the Colonial Dames before signing on to the project, it was undeniably Evangeline Andrews, past president of the Connecticut Society from 1927 to 1933, who guaranteed the architect's final stage of work would be executed by seeking funding from a New Deal work relief agency.

The project faced a minor hurdle, however, when the federal government transferred the work relief program from the FERA to the newly established WPA in the summer of 1935. Relief administrator Little told the board that this change made it impossible for the CERC "to assume any responsibility regarding this project," and she suggested it take the matter up with Matthew A. Daly, the recently appointed WPA state administrator and former Democratic state senator.[66] Andrews wrote directly to the "gentlemen" of the WPA to implore them to fund the necessary restoration work, explaining that "because of the general depression, the trustees are unable to procure, as in the old days, sums from individuals who are interested in the preservation of this fine old house." Andrews referred to customary sponsors from the private sector whose financial contributions withered during the years of economic decline. For example, throughout the summer, she sent letters asking for contributions to members of the Connecticut Society, but the response had been disappointing; members pledged only $853.50 through this campaign.[67] She astutely recognized that while traditional sources of funding dwindled, New Deal programs could provide federal monies for civic projects.

The transition from the CERC to the WPA, which Andrews described as "a blow at first" because of the project's uncertain status, turned out to be extremely advantageous for the Board of Trustees. The CERC would have required the board as project sponsor to cover the cost of the architect and materials, amounting to $12,365, to match the agency's contribution for labor. However, the WPA, if it approved the project, would pay for both the materials *and* labor. Since the board demanded its own architect—J. Frederick Kelly—be named to the job rather than pulled from the local relief register, it would be responsible for providing only the architect's fee of $2,365.19.[68] To preemptively raise money to cover his commission, the board sent a letter requesting contributions to SPNEA's corresponding secretary and Kelly's friend William Sumner Appleton. Appleton replied to Andrews with a guarantee of

$1,000 and added that should the WPA funding materialize, it would be "a highly creditable performance for our Society."[69] The organization's secretary undoubtedly wanted SPNEA to claim some credit for the preservation of the house if the restoration as a WPA venture came to fruition.[70]

Andrews excitedly wrote to trustee Frederick C. Norton about the prospect of the WPA taking on the restoration, describing the potential federal project as "something that will really put the house on the map and make it a very valuable contribution to Connecticut's history and cultural welfare."[71] Andrews optimistically viewed the Whitfield House becoming a New Deal project as a way to acquire federal resources for the promotion of the museum as a tourist attraction. Hopeful to secure funding for the final stage of restoration at last, the Board of Trustees filed a WPA application on November 4, 1935; a week later, relief administrator Little forwarded it to the central DC office.[72] President Roosevelt approved the project on November 26, 1935, and ten days after that the U.S. comptroller general sanctioned the full amount requested of $23,650. The trustees shortly afterward received the good news and promptly commenced preparations for the onsite work.[73] The Henry Whitfield State Museum quickly closed its doors to the public in mid-December, and staff relocated contents of the museum to the curator's house on the premises. Some pieces of rare colonial furniture were transferred to the New Haven Colony Historical Society, which agreed to store and exhibit items while the museum remained closed. The trustees placed the museum's silver pieces in the vault of the bank in Guilford and stored some books in the Guilford Free Library.[74]

Once the museum's collections were dispersed for safekeeping, the WPA project officially began at the house on December 28, 1935, with an initial assignment of ten men, including newly named superintendent Frank Spencer.[75] With this labor force, Kelly began to enact his long-prepared plan, overhauling both the exterior and interior of the Old Stone House to first finish undoing Isham's work and then to recreate the 1640 structure. First, Kelly lowered the walls two and a half feet to their original fifteen-foot height, and then he increased to its original sixty-degree pitch a new gabled roof, which he covered with split, hand-shaved cypress shingles. Kelly also introduced dormers to the roof (as seen in fig. 7) because excavations of the fireplaces at either end of the house led him to conclude that windows could not have been placed on the gable ends (north and south walls) owing to the large size and location of the original chimneys. At Kelly's direction, WPA workers installed new casement windows with diamond-shaped quarrels of glass set in lead bars, which were reproductions of seventeenth-century-style windows. The frames were recessed at an angle of about thirty degrees, determined by the discovery of an impression in the clay mortar on one of the jambs made by an original window frame. Joe Lynch, a local painter and "restorer of old buildings" assigned to the WPA project from the relief register, salvaged the double-hung sash windows with diamond panes installed in 1903 by Isham for use in another Guilford house.[76]

Inside the museum, workers removed the interior finish from Isham's imaginative two-story Great Hall, including the paneled oak chimney breasts at each end of the room, oak wainscot and ceiling beams, and the oak staircase in the eastern side of the room.[77] When restoring the north wall and its large chimney flue, Kelly made what he called "the great discovery of the restoration." He found vestiges of original masonry of a second-floor fireplace, definitively answering in the affirmative the long-debated question of whether the main part of the house had a second story or was open from floor to garret. Further meticulous work revealed a blackened, plastered lining of clay containing hay or straw and the seat of the original fireplace lintel, measuring fourteen inches in height by ten inches in depth. Kelly determined the lintel to be originally of timber because an impression in the old clay mortar showed the grain of wood. In the angle of the west and south walls of the second floor Kelly reintroduced an embrasure—a corner window about a foot wide built for the firing of a cannon—that, though he considered an "unlikely place for one," appeared in Smith's plans of the house, which he closely followed (fig. 10).[78]

Kelly likewise used Smith's measurements to alter the house's interior dimensions, modifying the first story to a height of 7-2/3 feet and the second story to 6-3/4 feet.[79] He incorporated moveable partitions in the first floor, like Isham had done, partly because they had "long been a tradition connected with the house" but also because he thought it likely the space had a divider since the minister's home in 1640 also functioned as the town meeting hall and first church of Guilford. The partitions, called baffles, were hinged walls made from vertical feather-edged boards that hung side by side from the ceiling. Turning small wooden cleats released the baffles, which swung down into a vertical position. This effectively partitioned the room and prevented cross drafts between the north and south fireplaces, more efficiently heating the space in cold New England seasons.[80] All of the oak framing timbers for the roof and window lintels were of native white oak cut from

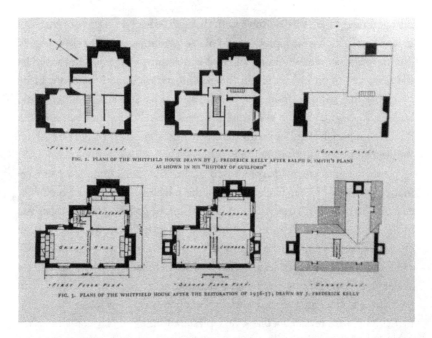

FIG. 2. PLANS OF THE WHITFIELD HOUSE DRAWN BY J. FREDERICK KELLY AFTER RALPH D. SMITH'S PLANS
AS SHOWN IN HIS "HISTORY OF GUILFORD"

FIG. 3. PLANS OF THE WHITFIELD HOUSE AFTER THE RESTORATION OF 1936-37, DRAWN BY J. FREDERICK KELLY

FIGURE 10: J. Frederick Kelly's plans for the restoration of the Whitfield House. Source: "Restoration of the Henry Whitfield House, Guilford, Connecticut," *Old-Time New England* 29, no. 3 (January 1939): 78.

local forests by the Bartlett sawmill in North Guilford and given an adzed finish. Many of the materials used in the restoration were salvaged from other historic houses. The oak flooring of the first floor was from an old house in Scotland, Connecticut, and the white pine and whitewood feather-edged boards of the moveable partitions were brought from historic structures in East Windsor and Bolton.[81] Door hinges and latches also were salvaged from old Connecticut houses, but Kelly hired the McDermott Company of West Haven to hand-forge from Swedish wrought iron the hinges, stays, and fasteners for the casement windows, which were copied from seventeenth-century English designs.[82]

In undoing Isham's work, relying on historical documents, and using local native and salvaged materials, Kelly felt he produced a more truthful representation of seventeenth-century colonial stone architecture. Like architects Albert Simons and Samuel Lapham, in charge of Charleston's Dock Street Theatre restoration, Kelly believed salvaged material from nearby historic homes authenticated his work, imbuing the state museum with the history embedded in real historical objects from others buildings of similar age. As Simons and Lapham argued that repurposing architectural elements from an antebellum mansion helped recapture planation society, Kelly likewise viewed the domestic elements relocated to the Whitfield House as crucial pieces in recreating an appropriate Puritan home. In truth, Kelly's choice did not differ altogether from his predecessor Isham's professed goal to recreate the "flavor" of the seventeenth century. Yet, there *was* a difference in the use of old materials at the Whitfield House and in the Dock Street Theatre project. The southern architects employed residential architectural elements to fabricate a long-demolished antebellum theater, appropriating domestic architecture for a commercial space. Kelly utilized domestic material in the restoration of an educational institution tasked with teaching visitors about Connecticut's long history; thus, the material from old homes on display in the museum contributed to the presentation and interpretation of the state's historical periods, although the salvage efforts were not contextualized as such. Therefore, while still taking creative liberties, Kelly's use of repurposed materials was more aligned with institutional mission than that of his colleagues in the South.

Although the focus of the WPA project was on the house's structure, the Board of Trustees attended to the broader cultural landscape of the historic site, considering both its architecture and natural features. While Kelly worked on the house, the board retained the services of professional landscape architect Beatrix Farrand, supervisor of landscape gardening at Princeton University since 1915 and consulting landscape architect at Yale University since 1923. Farrand's association with the Whitfield House began about a year before the start of the WPA's restoration when she met with Evangeline Andrews to offer suggestions regarding the landscaping of the property.[83] The following year, as Kelly's plans unfolded, Farrand revisited the site and made new and revised suggestions. They included building low stone walls surrounding the house, creating a garden enclosure to the northwest of the house, and planting flora native to both Guilford and southeast England.[84] The WPA funding covered the materials and labor to create Farrand's imagined seventeenth-century landscape, but the Board of Trustees itself paid for Farrand's professional services, as well as those of Grafton Peberdy, landscape architect at Yale University, who helped execute Farrand's vision and oversaw the construction of the stone walls and the planting of approximately 565 trees.[85] Like other professionals in art, literature, architecture, and associated fields, Farrand astutely utilized New Deal programs to build a diverse port-folio of public works and advance her career, which blossomed in the decades following the Depression.[86] Both Kelly's and Farrand's appointments to the WPA project at the Whitfield House reflected new employment opportuni-ties available to white-collar professionals during the New Deal: institutional work from government agencies, rather than private commissions.

"Faulty Work" at the Old Stone House: Material Delays, Labor Problems, and Local Criticism

Despite years of planning and the aid of professional expertise, the work at the Whitfield House progressed slowly and haltingly, much to J. Frederick Kelly's and the Board of Trustees' chagrin. The most consistent and serious prob-lem Kelly and WPA superintendent Frank Spencer faced was the shortage of building materials. The site often lacked the necessary amount of cement, lime, and timber, which prevented the masons and carpenters from doing their jobs. Since the masons could not work on the walls, the carpenters could not begin to build the roof, so the house remained roofless and susceptible to the elements for most of the spring of 1936 (fig. 11).[87] In mid-June, Kelly complained of the problem to WPA state administrator Matthew A. Daly. After their exchange, Kelly received a shipment of some building materials, but by the end of July the workers were once again out of lime and cement. Frederick Norton, as president of the Board of Trustees, wrote to Daly "to beg of you to help," and explained to Governor Wilbur Cross that the board was "in despair about the impasse at the Whitfield House."[88] Kelly's and Norton's efforts were temporarily effective; immediately upon receiving Norton's letter, Cross called Daly, who assured him that delivery of materials would be has-tened.[89] But despite this pledge, materials continued to arrive behind sched-ule. Kelly wrote again to the state administrator that the constant delays not only hindered the progress of the project but brought "the whole organization of the WPA into bad repute as well." If conditions did not improve, Kelly threatened to refer the situation to the WPA's central office in DC.[90]

Kelly's poor estimation of project superintendent Frank Spencer exacer-bated his frustration with the relief agency. The architect complained that Spencer failed to put in requests for materials when asked, concluding that the "wretchedly slow rate of progress" was "due as much to Spencer's lack

FIGURE 11: WPA workers rebuild the roof of the Henry Whitfield House. Source: Henry Whitfield House Restoration and Landscaping Projects, 1900–1940, RG024:001, State Archives, Connecticut State Library, Hartford.

of efficiency in this respect" as to the WPA's inability to provide materials in a timely manner. Other frustrations with the superintendent included Spencer's unexplained absences from the jobsite, his "inexcusable blunder" of improperly installing the stair handrail, and his failure to give adequate directions to the workers, which resulted in having to redo a steamfitter's work.[91]

Kelly's problem acquiring materials and his judgment of Spencer as a subpar superintendent heightened his irritation with the bureaucracy of the WPA. For example, he bemoaned the difficulty of hiring skilled laborers for particular jobs. The WPA required project sponsors to employ men from the local relief register, so carpenters, stone masons, painters, and other unemployed laborers from Guilford and nearby towns were hired at the Old Stone House.[92] This protocol created problems, however, when the complicated restoration required more expert skill. When the work started in December 1935, WPA regulations permitted 10 percent of non-relief labor, but a later federal ruling cut down the amount to 5 percent so that when a steamfitter was needed in November 1936, the Whitfield House project already employed the maximum non-relief labor force.[93] By then, Matthew Daly had resigned as WPA state administrator to serve as senator in the Connecticut legislature, so the Board of Trustees had to petition the new state administrator, Robert A. Hurley, to allow an exception for the hiring of a steamfitter.[94] The next month, however, the project's stonecutter was pulled off the job because a government Social Service investigator discovered that he had a small amount of life insurance, making him ineligible for tax relief. Another stonecutter was not found until the end of the month, prolonging the completion of this specialized work.[95]

In addition to navigating obscure hiring practices, Kelly became vexed at times with the relief workers themselves. Throughout the project, he expressed contradictory views, both criticizing and praising the men on the job. In a progress report to Evangeline Andrews, Kelly wrote, "It is to be deeply regretted that we have been denied the services of better qualified men." Yet, in the same report he admitted that "the spirit and willingness to work of all men on the job has been all that might be desired. The trouble has been, therefore, not in the men themselves but in the system of appointment."[96] He voiced his feelings more candidly in the personal journal he kept during the project: "The two Italian masons appear to be the only ones who work, whether they are watched or not. The Yankee carpenters are the men most inclined to loaf."[97] Despite relying on what he considered an inferior labor force, Kelly remained confident that the project could be completed successfully. Over time, his views toward the workers softened, perhaps reflecting sympathy for the men's status as relief workers.

Regardless of his feelings toward the laborers, Kelly's annoyance with the complexity of WPA bureaucracy persisted. By late summer of 1936 he admitted to Andrews that he was "so fed up with this noble organization [the WPA]."[98] The kind of labor issues Kelly dealt with were not unique to the Whitfield House project by any means; most WPA projects experienced setbacks and had to adjust as federal relief regulations changed, which they did often. Still, neither Kelly nor most Guilford residents likely followed the intricate workings of the ever-evolving bureaucratic system of the work relief program. Moreover, Kelly's professional experience included primarily privately commissioned work, and he likely found it vexing to be bound by the WPA's specific rules and to work with men he did not personally hire nor directly supervise when finally executing his long-awaited restoration.

Further complicating Kelly's and Andrews's rocky relationship with the WPA were external forces, chiefly unexpected local criticism. Kelly suspected "some hidden opposition to the project," and Board of Trustees president Frederick Calvin Norton attributed the difficulties with the WPA to "some political disaffection here in Guilford itself."[99] Local antagonism toward the project was brought to the board's attention by Dr. Walter Steiner, the grandson of Guilford historian Ralph Smith, whose nineteenth-century plans of the house guided Kelly's design, and a member of both the Board of Trustees and the Restoration Committee. Steiner wrote to Frederick Norton in the midst of the difficulties Kelly and his team faced acquiring building materials, explaining that while visiting the Whitfield House the previous day he "learned from an old friend that the men had done three days' work in two months." Moreover, this unnamed friend told Steiner that the many tourists who traveled to Guilford to see the house were disappointed to find it in the process of being restored with a sign of "Not Open Yet" on the door. To Steiner, this sign left "a bad impression" on the visitors.[100] When Norton telephoned Kelly to describe Steiner's attitude, the architect placed the blame on the WPA, excusing the Restoration Committee and himself from any responsibility for the slow progress and calling this friend of Steiner "guilty of an absolute misstatement of fact."[101] Yet, Steiner's friend was not false in their observations. At the time of the visit in question, the government recently had reduced the number of hours per month WPA laborers could work in order to meet the demands of labor unions and pay higher wages.

Evangeline Andrews, as president of the Restoration Committee, responded to Steiner's "uncalled for and unfair" letter. In her reply, Andrews referred to the political dissatisfaction of unemployed workers in Guilford, but she would not have had an intimate understanding of local labor politics, nor would Kelly or the other trustees. They were all well-to-do professionals, and many did not call Guilford home but rather lived in other Connecticut towns. Nonetheless, Andrews attempted to explain the situation to Steiner. Guilford, she enlightened him, "has politicians who pull wires, and an uncontrolled and unpunished group of gangsters who are allowed to go to Mr. [Frederick] Norton's house at night, break his windows, and throw eggs through his windows, all because he will not employ these ruffians to do the work on the house."[102] Properly admonished, Steiner replied to Andrews with sympathy for the board and its trouble with the WPA, expressing his "disgust and resentment" at the poor treatment directed at Norton by Guilford's workers on relief. Despite his acknowledgment of the limitations the WPA placed on hiring practices, Steiner still thought "it well if possible to have the kindly feelings of Guilford residents in this matter."[103]

In his communication with Andrews, Steiner referred to others in town who disapproved of the Whitfield House restoration, one of whom was Eva

Bishop Leete. Leete was a member of the Connecticut Society, founder of the local Dorothy Whitfield Historical Society, and longtime president of the E. B. Leete Company, which earned her local recognition as an expert dealer in colonial antiques.[104] When Kelly and Norton visited the house in July 1936, museum curator Ruth Lee Baldwin informed them that Leete had recently led to the house a group of women, many likely members of both the Dorothy Whitfield Historical Society and Connecticut Society, who were "outspoken in their criticisms." The appearance of the new dormer windows in Kelly's steeper roof and its covering of handmade cypress shingles especially affronted them (the new roof can be seen in fig. 7). The group also made comments like, "it was never there before," referring to the stone wall surrounding the property constructed by the WPA; "it has always been there," in response to the removal of a cement walk leading to the house; and "the roof was all right as it was," regarding the decision to replace the roof with one of historically accurate pitch. Leete told Baldwin that Kelly and the WPA "had absolutely no authority" for making such changes to the house.[105]

These women believed that Norman Isham's restoration two decades prior had resulted in an accurate reproduction of the house, and they worried that the WPA laborers, under Kelly's and the board's guiding hands, were actively destroying its historical integrity. To them, the house and the history enshrined within were sacrosanct as a site of civil religion, and any alteration weakened its authenticity and therefore power as an educative force. But public memory and historical accuracy are frequently in conflict. These local memory keepers who voiced criticism of Kelly's re-restoration preferred preservation work that reflected their own imagined historical past rather than one based on documentation. To them, altering the house from what they regarded as its most authentic iteration, despite Isham openly recognizing that the 1640 house no longer existed, did a disservice to visitors expecting to encounter the real Puritan past within the museum's walls. The Colonial Dames still believed they should exert influence over local matters of heritage and cultural preservation, especially regarding the maintenance of historic homes, and they openly challenged Kelly's work.[106]

Leete and her friends intended to implore the state chapter of the Colonial Dames in Hartford to have the work at the museum stopped immediately. She conveyed to Norton the opinion that the town "looks with profound regret upon the faulty work which is now being done." She even threatened to call a town meeting "to bring the desecration to a halt, and save the house if possible, before it is torn to pieces and ruined."[107] Kelly's earlier apprehension, voiced in 1930, that the Colonial Dames' involvement at the house would be contentious proved prescient. It is important to note that the nature of the relationship between the Board of Trustees and the Connecticut Society regarding the management of the state museum throughout the 1930s is unclear. The museum remained "under the auspices" of the Connecticut Society, but the trustees had legal authority over the institution. What *is* clear is that the society was divided in its feelings about the changes made at the Whitfield House: Evangeline Andrews was in full support as head of the museum's Restoration Committee, but some members, like Leete, vehemently disagreed with Kelly's plan and were not directly involved in decision-making during the WPA project.[108]

Whatever the cause, or causes, of the local criticism, Norton and Kelly had to bear the brunt of it. Kelly described the townspeople's attitude as "most disheartening." His attention to the use of appropriate materials and building techniques to make the house resemble as closely as possible its original 1640 condition seemed to have been wasted.[109] Kelly conjectured that Leete's and other Guilfordians' adversarial attitude revealed "perhaps not so much a definite disapproval of our work on technical grounds, as . . . a crystallization of sentiment against the W.P.A. . . . and the New Deal in general." As proof, he argued, no one objected to the scale model of the house he made when it was displayed in the Guilford Free Library in the summer of 1935. The criticism, he felt, was born out of recent dissatisfaction rather than longstanding resistance to his designs.[110] There was some truth to Kelly's defense. Residents' disapproval of the project did not surface until a year after his plan had circulated, and besides the women's pointed remarks about architectural features, local criticism centered on the involvement of big government in local affairs. Like most small towns in Connecticut, a state that had recently transitioned to Democratic control, Guilford residents expressed varying positions on the efficacy of federally administered work relief. In a November 1837 "Post Card Forum" on the question of unemployment relief in the local *Shore Line Times*, one writer called for taking relief "out of the hands of an expensive Federal Bureaucracy" and putting it back in the hands of local administrators "who intimately know the town and its residents . . . [and] will put an end to the useless waste of tax-payers' money."[111] This resident's opinion reflected the disdain for any wasteful projects, or "boondoggles," as well as the penchant for local governance without federal interference. In truth, local discontent probably derived from a combination of disgruntlement with government relief administration *and* umbrage at women's minimized role as custodians of the Whitfield House.

In the fall of 1936, the board started a campaign to convince residents of the value of the project and defend Kelly's professional decisions. Curator Ruth Baldwin and WPA superintendent Frank Spencer began passing out pamphlets to visitors that explained the WPA work to assuage ill will toward the project. As a result, Baldwin thought, there was "a much better understanding of the reasons for the restoration."[112] To respond publicly and directly to the criticism of both Kelly as architect and the WPA's involvement with the site, Evangeline Andrews published a lengthy letter in the *Shore Line Times* in early December 1936. She urged folks to read about the history and restoration of the museum in the new pamphlet and challenged those who continued to find fault with the project to "bring forward specific criticisms and specific proof, historical and architectural, of the same quality and scholarly value as that we offer." Recognizing that some disapproval resulted from the association of the site with a New Deal agency, Andrews explained that *only* the federal government could have provided sufficient funding during the Depression, as fundraising campaigns from private sources had not raised enough money. However, she made sure to mention outside grants from the Connecticut Tercentenary Commission, SPNEA, and private individuals across the state to demonstrate the project had moral, if not financial, support from its traditional backers. Meanwhile, the services offered by the Art School of Yale University and Beatrix Farrand were "generous gifts in themselves" and indicated the endorsement of leading experts in art history and landscape architecture.[113] Finally, Andrews's letter assured Guilford residents that rather than destroying the historic integrity of the town's most beloved shrine, the WPA project honored the Whitfield House's first occupant by resurrecting the structure's original design and appearance. It is unclear what immediate effect the letter had on residents, but board members did not receive any more injurious messages or incur damage to their persons or property following its publication.

The Puritan Past Restored at the State Museum

While the trustees and Kelly faced difficulties in Guilford, neither the state nor the central WPA offices recognized any problems with the work underway at the Whitfield House. In the spring of 1937, Connecticut state administrator Robert A. Hurley received a follow-up letter from the WPA's central office asking for potential projects that could be considered "historic shrines." Julius Stone, the associate director of the Division of Information, had read an article Hurley published in the *Hartford Courant* that described the restoration of both the Whitfield House in Guilford and the Old Town Hall in Fairfield. Stone thought both projects "quite evidently . . . come under the description of restoration of historic shrines."[114] The state WPA office considered Guilford's Old Stone House one of the state's most successful projects funded by the agency and featured it on the cover of the January 1937 edition of the *Connecticut Work in Progress* magazine, a monthly publication that broadcast news of WPA work in operation throughout the state.[115]

Additionally, an internal WPA progress report on the Whitfield House described the federal agency as "agreeing with the Guilford citizens that the little building was much too important historically to be allowed to disintegrate." The WPA "stepped forward in the role of savior" to protect "the oldest house in Connecticut, and one of the oldest buildings in the country." Indeed, WPA publicity played up the supposedly poor condition of the Old Stone House before it was rescued through the work relief program. "Three centuries of sun, rain, and wind had gotten in their work, cracking and wearing it." Under the WPA's care, the "little house . . . about to crumble away" was returned to the appearance it held when the minister Henry Whitfield "prayed for the souls of the faithful and tried to convert the heathen savages." The WPA report both praised the agency's intervention and used colonialist language to support the revival of a Puritan past.[116]

In the midst of architect Kelly's struggles with WPA labor and materials, the Board of Trustees' campaign to win local residents' approval, and the WPA's broadcasting of the project as an unqualified success, the Henry Whitfield State Museum faced a fresh, unexpected challenge leveled by the state government. In 1935, as part of his reform platform, Democratic governor Wilbur Cross appointed a Commission on the Reorganization of State Departments to make legislative recommendations for streamlining the state government, which at the time consisted of 160 agencies. By 1937, Cross had the support of a Democratic General Assembly to enact the reorganization scheme, which proposed placing the Henry Whitfield State Museum under the Department of Parks and Forests and abolishing the thirty-year-old Board of Trustees.[117] The board surely felt blindsided by this proposal after the unanticipated and fraught uphill battles it already had waged to see Kelly's restoration succeed. The board vehemently protested the plan, as Evangeline Andrews feared that the Old Stone House would "lose its identity and become a unit of a chain of state institutions . . . in a manner closely resembling the modern chain store."[118] She worried that the imposition of increased state control over local landmarks would result in sterile management typical of modern bureaucracy. She also expressed her dismay that the state legislature considered removing the property from the responsibility of those possessing the local knowledge and affection to make the house "the distinguished living museum that we have been working for all these years."[119]

Backing the board's position against the reorganization plan, hereditary and social organizations whose patronage supported previous preservation work at the house rallied in protest. Representatives from the Connecticut Society, the Sons of the American Revolution, the Daughters of the American Revolution, and the Dorothy Whitfield Historical Society of Guilford sent letters to Cross expressing their opposition.[120] Even Dr. Walter Steiner, who had stirred the pot of local criticism just six months prior, called the plan to transfer the state museum to the Department of Parks and Forests "really most suicidal" and "criminal."[121] The visitor objections Steiner had brought to the board's attention centered on architecture and aesthetics; more egregious was the proposal to remove control of the institution from a local governing body, which he came to believe oversaw the site with admirable intentions.

Guilford townspeople's fierce reaction to the state plan demonstrated the widespread credence that stewardship of historic properties was a civic duty vested in local control. In a letter to Cross asking him to exempt the Whitfield House from the reorganization plan, one museum trustee argued that "the sense of personal possession that rightly exists in Guilford toward this property, which is associated with the earliest history of the town and which is a valuable socializing influence, will be lost to those who feel it most."[122] This argument addressed the important community building work that occurred when caring for a beloved local historic landmark. It also illustrated Guilford residents' proprietorial attitude toward their oldest structure; they had the privilege to criticize the restoration work, but outsiders, including the state government (despite owning the house), did not have the right to remove the property's care from the people justly responsible for maintaining its legacy.

In a well-earned victory after many trials and tribulations, the local *Shore Line Times* announced that Connecticut's General Assembly voted against the reorganization plan on April 8, 1937. The same week that the board celebrated this news, Kelly and his team of WPA workers completed the restoration of the Whitfield House, a remarkable achievement considering the logistical and labor challenges, the negative views of some townsfolk, and the threat of the state reorganization scheme.[123] After almost sixteen months of being closed to the public and a total federal expenditure of $20,046.44 for labor and materials, the Henry Whitfield State Museum reopened on April 26, 1937, with plans to hold a formal celebration in the fall.[124] The staff retrieved from their temporary holding places the museum's collections to which they added seventeen pieces of rare seventeenth-century furniture on loan from the impressive Mabel Brady Garvan collection at Yale University.[125] The board named as assistant curator trustee Annie Jennings, who quickly offered to finance a publication that documented the restoration. The specific purpose of the book was to stir up interest and boost visitation to the museum by making available photographs and Kelly's personal notes he had kept during the project.[126] Over the summer, a photographer from the National Geographic Society traveled to Guilford to capture images of the finished project, and the curator and historic home restorationist Joseph Downs from the Metropolitan Museum of Art visited and commended Kelly for his excellent work.[127]

Meanwhile, the Board of Trustees, saved from dissolution, prepared for the official commemoration celebrating the museum's reopening on October 20, 1937, exactly forty years after the state acquired the property. Speakers at the opening ceremony included prominent politicians, historians, and preservationists.[128] After twenty-seven years as corresponding secretary of SPNEA, William Sumner Appleton remarked that "in all of New England there has been no more notable example of the preservation of an historical antiquity" than that of the Whitfield House.[129] When dedicating the site, Evangeline Andrews thanked the many players involved in the restorations of the Old Stone House over the years, specifically naming the Connecticut Society, architects Norman Isham and J. Frederick Kelly, current and former trustees, the Tercentenary Commission of Connecticut, Yale University, Beatrix Farrand, and SPNEA. She also conceded that the historic shrine's restoration would not have been financially feasible "without the co-operation of the Federal government," although she admitted that the trustees and Kelly had struggled to navigate the WPA's bureaucracy.[130] In her address, Andrews overstated the supposed political crises aroused by the involvement of the WPA, which continued to view the Whitfield House restoration as an outstanding success. As evidence, in 1940 the director of the Division of Information in the state WPA office selected the Whitfield House as one of the top ten most interesting projects in Connecticut, writing that "as restored this famed old house stands as a credit to the WPA."[131]

In her speech, Andrews voiced her hope that the historic structure would help subsequent generations resist modernizing influences, such as urbanism and consumerism, a viewpoint that expressed an antimodern historical consciousness:

> Perhaps, if in the future, life in our country becomes increasingly materialistic; if in our towns the ubiquitous chain-store and other standardized and ugly units of modern buildings shoulder out the simple old houses which with their gardens make for pleasant and friendly living; if cities and towns lose their old trees and open spaces; then perhaps this old Whitfield house, standing foursquare to the winds and surrounded by a generous acreage and the kind of trees that might have been its companions three hundred years ago— perhaps then it may perform for Guilford and the State a service not dreamed of today.[132]

Andrews considered the Whitfield House a fortification of community, a modern defensive structure not against potential Indigenous attacks as in 1640 but against forces that diminished local character, reduced space for living and enjoyment, and catered to commercialization and materialism. In the 1930s, the restoration of the Old Stone House both preserved Guilford's historical identity and stood as a bulwark against unwanted change, which explains why the most vociferous reactions came from members of patriotic organizations that represented the traditionalism of the Puritan Yankee. For residents, the house was a sacred historical text that people could interpret to learn lessons about the past, and they responded negatively to Kelly's changes that seemed to modify its architectural language. At the museum's opening ceremony, Samuel H. Fisher, chair of the Tercentenary Commission, reasoned that while schools taught children about the state's history, it was through the architectural preservation of landmarks like the Whitfield House that they might "absorb something of the early life of our people."[133]

WPA literature about Connecticut similarly promoted the idea that architecture embodied social values and offered a sense of permanence. The American Guide Series' *Connecticut: A Guide to Its Roads, Lore, and People* (1938) described the state's architecture as "the most permanent expression of its social life—the translation of habits of life and modes of thought into wood and stone." The Connecticut colony presented to its earliest settlers a "struggle for existence . . . [that] produced a simple and sturdy indigenous mode of building less influenced by foreign precedent than any other Colonial architecture."[134] The solid, immutable stone of the Whitfield House encapsulated the qualities associated with the Puritans in New England lore, repeated in the American Guides, and reiterated their reputation as an inflexible and unyielding breed of people to match their enduring architecture. The state guidebook, moreover, described the quarried stone of Guilford used to build the Whitfield House as the same source from which emanated sites of national historical significance, including the Statue of Liberty, the Brooklyn Bridge, and Lighthouse Point in New Haven.[135] The association between the local stone and places representative of freedom, engineering innovation, and security linked Guilford with broader developments in the American experience and proffered the town as a foundational center of the nation. Perhaps that is why Board of Trustees president Frederick Norton proudly proclaimed that "this solid gabled structure is to us Guilford folks on a par with the State House at Philadelphia, or Mount Vernon near Washington."[136]

Within seven months of opening, over five thousand visitors representing forty-two states and other territories and foreign countries had crossed the threshold of Connecticut's state museum.[137] As the museum enjoyed success, the town began preparations for its tercentenary celebration to be held in September 1939, commemorating the three-hundred-year anniversary of the arrival of Henry Whitfield's company. Defense of the town's Puritan forebearers was the major theme underlying the festivities, with speeches from leading townspeople titled "The Spirit of the Puritan Pioneers" and "Puritan Contributions to the Life of Today."[138] Like other public historical reenactments of the 1930s, Guilford's tercentenary included a pageant, titled "Heart of America," which depicted Whitfield and his followers' settling of the town.[139] The pageant's finale concluded with a foreboding yet optimistic declaration, reminding the audience of their own economic and political moment of crisis: "We are caught in the treacherous currents of terrifying change. For us there are no frontiers, no new continents to which we can escape. We must stand where we are and face squarely the colossal problems of our time. But let us face them armed with the inspiration which the Past can give."[140] Both the Henry Whitfield State Museum's restoration and the tercentenary commemorations that followed reflected commitment to a local mythology about the Puritan founding of Guilford which profoundly shaped how residents imagined themselves. In the 1930s, Guilfordians experiencing the economic turmoil of the Depression and political shifts in state politics perceived the built environment of early Connecticut—and their own Old Stone House especially—as an expression of the rigidness, hardiness, resourcefulness, and enduring colonizing success of the Puritans, a historical origin story that could fortify their community against future political and social upheavals. In this manner, project leaders with support from the state government leveraged New Deal funds, though not without discord, to subtly and subversively through the historic built environment push back against the onslaught of modernizing forces.

CHAPTER 4

Cultivating the Minnesota Frontier

The Charles A. Lindbergh Boyhood Home and
State Park, Little Falls, Minnesota

Nine years ago today hundreds of visitors swarmed to the vacant Lindbergh homestead to view the boyhood scenes of the youth, who had just completed the first non-stop flight across the Atlantic. Hundreds swelled to thousands the next year—it was Sunday—and the house was in a state of delapidation by nightfall, so eager were souvenir hunters to take away a piece of it. Today, to make the anniversary more outstanding, word was received from WPA offices that final approval had been given the $23,777 Lindbergh State park development project, which is sponsored by the State Park commission.

"Lindbergh Park Program Approved on Anniversary,"
Little Falls [MN] Weekly Transcript, May 21, 1936

Upon landing in France on May 21, 1927, after his nonstop solo flight from New York to Paris, twenty-five-year-old Charles A. Lindbergh became an instant celebrity, adored worldwide by fans starstruck by his bravery and mastery of flight. That day, his 1907 boyhood home in Little Falls, Minnesota, became a historic shrine despite being less than thirty years old. The *Saint Paul Pioneer Press* reported that before this awe-inspiring flight across the Atlantic, the "old Lindbergh residence" stood "unoccupied, neglected and weedy. . . . [It] drew no attention from passersby."[1] The *New York Times* aptly characterized the home's sudden rise to notoriety as "just a farmhouse until the boy who once played about on its floors achieved great fame. Overnight it became a prized place of historic significance."[2] Lindbergh's success brought the small town of Little Falls in central Minnesota, population five thousand, and the modest farmhouse where the young flier spent his childhood into the national spotlight.

The Lindbergh home sat on property just southwest of the town center that the aviator's father, Congressman Charles Lindbergh, had purchased in 1898. When a fire in 1905 destroyed the first home Lindbergh Sr. had built for his family, he and his business partner Carl Bolander, "a kind of architect," constructed a second, smaller home on the same foundation using native materials.[3] The new story-and-a-half farmhouse, built between 1906 and 1907, sloped in the rear toward the Mississippi River to take advantage of the scenic vista. It had an exterior of rough-sawn lumber, beveled siding, plaster-on-wood interiors, and maple flooring.[4] The home's weatherboard cladding presented a familial and plain appearance with its light gray with white trim paint, stone foundation, gabled hip roof with red brick chimney, and simple front porch (fig. 12). When Anne Morrow Lindbergh saw her famous husband's boyhood home for the first time in July 1935, she described it as unremarkable: "a good-sized house, clapboard, of no particular form or style."[5] Folks in the area who frequented the Lindbergh abode called it "the queerest house in Minnesota" because of its center-hall plan with seven doors, but to Charles's mother, Evangeline Land Lindbergh, it was a quaint, charming family home filled with memories of her son's happy childhood on the banks of the Mississippi.[6]

With Charles Lindbergh's meteoric rise to fame, the unassuming farmhouse became Minnesota's premier destination overnight, immediately attracting fans of America's newest celebrity. The *New York Times* acknowledged it was Lindbergh's "own action that directed attention to the farm where he had spent his boyhood" when he visited Little Falls in August following his groundbreaking flight, drawing the built environment of his childhood into the limelight.[7] Throughout the summer of 1927, souvenir hunters eager to claim a memento associated with the flier journeyed to Little Falls and damaged the structure of the house in their excitement. After four years of abuse to the property, town residents successfully pushed for the creation of a ninety-three-acre Lindbergh State Park in 1931 to protect the family's homestead and secure a state appropriation for its maintenance. Four years after that, in 1935, the Minnesota Department of Conservation submitted a

FIGURE 12: Charles A. Lindbergh House, c. 1936, west (front) and south elevations. Source: Historic American Buildings Survey MN-79, Library of Congress Prints and Photographs Division.

proposal to the newly created Works Progress Administration to return the modest home to its 1907 appearance and to further develop the recreational state park lands.

The WPA shrine restoration at the Lindbergh State Park, thus, consisted of two components: the first restored the farmhouse's exterior and interior to the period of Lindbergh's childhood, and the second cultivated a modern park landscape to encourage outdoor activity and support New Deal environmental efforts. The dual nature of the federal project advanced the state Department of Conservation's and the WPA's overarching goal to materialize the mythologized frontier of Lindbergh's youth as a recreational playground for Minnesotans and tourists to enjoy. The emphasis on the frontier in the park project complemented the work of other New Deal programs, particularly Federal One and the Treasury Department's Section of Fine Arts, which similarly romanticized the nation's agricultural history in murals, plays, and travel guides. As historian Barbara Melosh has argued, New Deal art propagandized American agrarian life and "made the farm an icon of an idealized social and moral landscape. . . . Frontiersman and farmer exemplified the democracy and promise of life close to the soil of a new land."[8]

Although historian Frederick Jackson Turner declared the U.S. frontier closed back in 1893, the concept remained a central symbol that connected the past to the present for the general American public in the 1930s.[9] The frontier, cultural critic Richard Slotkin posited, became "a term of ideological rather than geographical reference" that helped Americans interpret the changing world around them.[10] Historian Warren Susman, meanwhile, explained Americans' glorification of the frontier during the Depression years as a search for "a native epic . . . that extolled the virtues of extreme individualism, courage, recklessness, aloofness from social ties and obligations."[11] In the mid-1930s, nobody exemplified the virtues associated with the pioneer past more than the wholesome and hardy adventurer Charles Lindbergh, before the looming war in Europe led to the celebrity's controversial status as a prominent spokesperson for the America First Committee.

Indeed, during the early years of the Depression, fans heralded Lindbergh as the ideal American man; the image of the young explorer earning his stripes by bravely conquering the newest frontier—the sky—abounded in the popular rhetoric surrounding his 1927 flight. The WPA project profited from Lindbergh's mythic status as a hero of America's heartland and helped curate an image of the flier as a modern-day pioneer. Describing the positive public reaction to Lindbergh's daring journey in the air, historian John William Ward wrote that the "lone eagle" symbolized "a long and vital tradition of individualism in the American experience."[12] Those who exalted the young adventurer likened him to other heroes of the American frontier like Davy Crockett and Daniel Boone, whose own birthplace in Birdsboro, Pennsylvania, was restored as a "national shrine" by the National Youth Administration with aid from the Boy Scouts.[13] Chief Scout executive James E. West wrote that Charles Lindbergh "called to the blood of the pioneer in every American boy," and the 1929 *Boy Scouts Handbook* featured Lindbergh's profile alongside Boone, Abraham Lincoln, and Theodore Roosevelt.[14]

Americans who admired and venerated Lindbergh believed he exerted a positive influence on the nation's youth by making the old-fashioned traits of hard work and modest living alluring in the "Roaring Twenties," a decade of perceived decadence and moral degeneracy. Historian Dixon Wecter, writing in 1941 about America's tradition of hero-worship, argued that Lindbergh enjoyed popular appeal because he was "a quiet rebuke to the Lost

Generation," a young cadre of American writers and thinkers whose experience of World War I led to their disillusionment and spiritual disconnection from traditional American ideals of patriotism and family.[15] In a tumultuous era of great social upheaval and changing economic fortunes, Lindbergh put forth an admirable image of a native son with a down-to-earth nature who achieved success and fame through grit and determination, an inheritance of the Upper Midwest's northern European pioneers not dissimilar to the legacy of the Puritans discussed in chapter 3. At a time when industrial developments drew large numbers of people from rural areas to burgeoning cities, and more and more Americans lost their direct connection with an agrarian lifestyle, Lindbergh "filled the desire for heroes built from common country stock" and "affirmed heartland values of self-reliance and independence."[16] These pioneer-like characteristics the aviator personified drew adoration from fans across the country and geographically pinned Minnesota—in particular, his small hometown of Little Falls—as America's moral center.

Yet, while Lindbergh embodied the historical spirit of a frontiersman charting a path in unfamiliar territory as he expertly navigated the skies, he also inspired hope in an industrial tomorrow. A paradox, even a conflict, between an agrarian past and an urbanized future thus resided in the New Deal project to remake Lindbergh's small family farm into a "natural" recreational area of the modern state park system. Imagery of New Deal programs also captured this dichotomy; for example, photographers of the Historical Section of the Farm Security Administration produced a "panoramic documentation of America life," which recorded, sometimes romantically, both the waning of the nation's agricultural era and the spread of urbanization in rural Americans' "march to the city."[17] American society mirrored this incongruity at large during the Depression as many people sought comfort in a fictionalized, uncomplicated rural past while they simultaneously and eagerly embraced modern industrial changes, many of these changes—or "improvements"—effected through federal relief programs.[18]

The Lindbergh State Park exemplified this tension between past and present, tradition and modernity, work and play. To produce the effect of a modern state park steeped in history, WPA workers built Rustic-style structures that evoked a specific Minnesota of yesteryear, a time when Lindbergh's own Swedish ancestors worked the land. Laborers adopted the building methods and rough-and-ready aesthetic of early frontier architecture to create a gentle and attractive site of leisure. At the same time, the project advanced new theories of park management and environmental conservation, which was a high priority in Minnesota's agricultural communities during the Depression years. In doing so, the built environment of the Lindbergh Park became "an expression of the romanticism of pioneer America" even as it endorsed the New Deal's modern conservation agenda.[19] The WPA's restoration of the Lindbergh homestead and its emphasis on celebrity, recreation, and environmental solutions endeared the project to Minnesotan families and farmers distressed by the dual economic and agricultural depressions. It reaffirmed the pioneers' laudable work ethic and love of nature, in the process emphasizing triumph over agricultural hardships and erasing the presence of Native history, and developed the state's recreational tourism while expanding the modern state park system.

Creating the Lindbergh State Park

Shortly after completing his transatlantic flight in May 1927, Charles Lindbergh embarked on a cross-country tour to promote aeronautics and greet his hundreds of thousands of fans. After visiting the Twin Cities of Saint Paul and Minneapolis, Lindbergh and his mother headed to Little Falls to visit the family's old homestead on August 25. Residents prepared for "the greatest day in its history" when Little Falls' "most famous native hero" would return to his roots.[20] A crowd of fifty thousand, about ten times the population of Little Falls, gathered to greet Lindbergh on his arrival. The celebratory homecoming included an elaborate parade with floats, bands and drums corps from several Minnesota towns, and replicas of Lindbergh's plane as well as the Statue of Liberty and the Eiffel Tower, monumental representatives of the cities where he began and ended his journey. The parade concluded at the city fairgrounds where the mayor, state representatives, and the governor gave speeches. The *Minneapolis Journal* reported that "the homecoming of Colonel Lindbergh to the soil of Minnesota is an affair of the heart."[21] From the earliest celebrations for the flier, it was clear that fans believed the physical place of Lindbergh's youth contributed to his personal achievements.

Before Lindbergh's arrival, the *Little Falls Transcript* published warnings that pickpockets and souvenir collectors followed the flier from city to city during his national tour. Some of Lindbergh's superfans—called "relic hunters" in the press—traveled to Little Falls to claim a tangible keepsake associated with the celebrity. Admirers did not want merely to catch a glimpse of the flier or gather printed memorabilia; they wanted to own a piece of the physical environment that had molded him. Martin Engstrom, longtime friend of the Lindberghs who had been keeping an eye on their house after they moved out of state, boarded the windows and padlocked the door immediately after the aviator landed in Paris. Despite these efforts, within a half hour Engstrom had received a call that people were loitering around the property, and soon enough fans ambushed the fairly secluded house.[22]

The city of Little Falls anticipated that the souvenir hunting would only escalate during Lindbergh's visit, so area police and National Guard units were assigned to the town for special duty in preparation.[23] The extra security proved ineffectual. The most aggressive of the thousands of tourists who visited the Lindbergh property that summer ripped boards from the sides of the house, chiseled pieces of rock from the foundation, carved their names and initials into the walls and ceilings, climbed old trees on the property, and dismantled Lindbergh's old Saxon automobile for its parts.[24] Figure 13 shows Lindbergh during his visit standing outside of the house, with scattered debris littering the ground and inscribed marks on the front porch post. The home's exterior presented a battered appearance, but so, too, did the interior: a beautiful mahogany cabinet stood stripped of its base and missing its glass doors, and an oak bookcase once holding Congressman Lindbergh's law books had been emptied. In their wake the ardent admirers left, according to a *Washington Post* article about the wreckage, "a dilapidated frame house."[25]

Little Falls residents were dismayed by the rundown state of the Lindbergh property, a place of immense local and now national pride. The Little Falls Board of Commerce wrote to Lindbergh's New York lawyer as early as October 1928 to express its dissatisfaction on behalf of the community with the management and lack of protection of the family homestead. The board representative explained that the home "is not in good shape and is in considerable need of repair," highlighting the major assaults perpetrated by tourists.[26] The board's concern expressed the deep affection Little Falls residents harbored for Lindbergh and his family's land, as well as the recognition that the property could become the city's foremost historic site and park—a boost to the local economy during sluggish years—if treated with care in proper proportion to the celebrity of its owner.

By 1931, the Lindbergh site had attracted thousands more visitors from across the nation who continued to damage the home and land. After over two

FIGURE 13: Charles Lindbergh (right) holding a "Little Falls" sign in front of his boyhood home, the Lindbergh House, August 25, 1927. Source: Minnesota Historical Society.

years of legal inactivity, prominent citizens and politicians with ties to Little Falls formed the Lindbergh Park Committee.[27] The Board of Commerce, on the recommendation of the committee, proposed a plan to Lindbergh and his mother to safeguard their homestead: the city of Little Falls would purchase the house and its acreage from the family, then "improve the property by restoring the home, cleaning up the entire property, and putting a care taker in charge . . . in other words making the whole estate into a State Park."[28] Nels Nelson Bergheim, estate attorney for the deceased Charles Lindbergh Sr., urged the family to pass the property over to the city to start the process of creating the park. For many years, Bergheim had rented out the farm for cash while the house itself remained vacant. The rent gave him "just enough to keep the house . . . in proper repair."[29] His appeal worked: Evangeline and Charles, along with his two half-sisters, ceded their interests and conveyed the 110-acre property by deed of trust to Martin Engstrom, their old family friend, with the understanding that most of the land was to be converted into a state park.[30]

While Lindbergh approved the creation of a recreational area to be enjoyed by the public, he was apprehensive to appear an attention-seeker. According to his personal lawyer, Lindbergh wished "that any attitude of his should not be interpreted as seeming to desire or encourage the establishment of any park or other institution as a memorial to himself or his actions."[31] At the time, Lindbergh remained one of America's favorite celebrities; Charles and his wife, Anne, still received over one hundred letters a day in 1931. The couple had trouble dining in public because they were so recognizable, and fashion magazines often featured the stylish Anne Morrow Lindbergh, a pilot in her own right.[32] Charles was a reclusive celebrity and he avoided the media the best he could, which made people all the more eager to learn of his activities. Engstrom assured Lindbergh he need not worry about people misunderstanding his motivation, as local desire for the park's establishment was obvious. Engstrom wrote to his friend that "our people are very anxious to preserve the place."[33] With final approval from the family secured, Engstrom made ready to transfer the property deed first to the city of Little Falls and then to the state of Minnesota, which would oversee its integration into the state park system.

As Lindbergh requested, the park technically was named for his congressman father who had died in 1924 of cancer while campaigning for the governorship of Minnesota on the Farmer-Labor Party ticket. Minnesotans, especially in rural areas like Little Falls's Morrison County, remembered Congressman Lindbergh for his staunch defense of farmers, his attack on trusts, and his opposition to American intervention in World War I.[34] A park named for him, and conveniently also for the famous son who shared his name, fit the political and cultural climate of the early 1930s. The Farmer-Labor Party, a coalition of farmers, organized labor, and small businesspeople, recently had gained control of the state with the election in 1930 of the party's first governor, Floyd B. Olson. Olson ended over fifteen years of Republican leadership and was reelected in 1932 and 1934, before dying while in office in 1936.[35] During his terms, despite initially facing a hostile legislature, he restructured the state government, secured greater unemployment benefits, and pushed for increased work relief projects. A strong advocate for farmers and agricultural reform, Olson was a committed conservationist, although personally opposing Roosevelt's 1936 domestic allotment plan for agriculture. He believed that commercial exploitation had "robbed our people of the greater part of their heritage of natural resources" and avowed to "guard what is left diligently and zealously." At his direction, the state legislature reorganized the Department of Conservation. The new plan placed Minnesota's state parks under the supervision of the Forestry Division and expanded the number and acreage of sites.[36]

The Lindbergh State Park was one of the first parks proposed after this reorganization, and Little Falls native Christian Rosenmeier, then serving as a Republican state senator, sought political support in the state legislature for its creation.[37] Former Republican congressman Ernest Lundeen, who by 1931 had joined the Farmer-Labor Party, also eagerly anticipated the formation of the park. He had "been fighting for this ever since 1925," two years *before* Lindbergh's famous flight, and gave at least twenty speeches on the radio in support of the park's establishment.[38] In early February 1931, Rosenmeier successfully pushed through the state senate the proposal to create a ninety-three-acre state park with an annual appropriation of $5,000 for maintenance and minor reconstruction. The Minnesota House Appropriations Committee passed the bill by unanimous vote, and with approval of both the senate and house, Governor Olson signed the bill creating the Lindbergh State Park during the first week of March 1931.[39] With this official news, the city of Little Falls announced that the "boyhood home of flier [is] to become [a] mecca of tourists."[40] The partnership of Republican and Farmer-Labor politicians to create the park set a precedent of political cooperation that later characterized the tenor of the WPA project to restore and expand the park. Lindbergh's hope that there would be "no opportunity . . . for anyone to take advantage of the situation—political or otherwise!" seemed to foretell the collaborative rather than conflictive interplay among the local, state, and federal entities involved in the park's development.[41]

In April, members of the Lindbergh Park Committee traveled from Little Falls to Saint Paul to deliver the property deed to Stafford King, the state auditor.[42] King, a Republican, was elected to his first term in 1931 at the same time Farmer-Laborite Olson entered the governor's office. King served on the state's Conservation Commission under the jurisdiction of the Forestry Division, which now supervised the state park system.[43] In May, the state auditor visited the newly created Lindbergh State Park and named as superintendent Martin Engstrom, who then appointed sixty-five-year-old Rufus Sutliff as caretaker. Sutliff was a longtime Little Falls resident and lived in the small tenant house across the road from the main house; the two homes were the only remaining dwellings on the almost one-hundred-acre park property.[44]

Preliminary plans for the park called for a speedy opening in the summer of 1931 with the main property fenced off and visitation regulated. King estimated that it would cost between $2,000 and $3,000 to "make the house presentable," so no immediate plans were made to furnish the interior as the annual $5,000 in state appropriations had to cover all park maintenance costs. However, the state auditor hoped that in time "trophies and relics can be secured" to outfit the home's interior.[45] That summer, the *Little Falls Daily Transcript* reported the restoration work to be "in full swing." Projects included rebuilding the porch overlooking the Mississippi River, replacing the exterior siding and missing foundation stones, and painting the house white to cover the thousands of tourists' names scribbled on the walls. The goal, according to Engstrom, was to make the house "look just like it did when the Lindbergh family lived there." Tending to the house was the priority at this time, but Engstrom, like King, had plans to expand the restoration. He sought to recover or reproduce some of the family furniture that went missing during the summer of 1927, and envisioned picnic grounds across the road from the main house.[46]

The creation of the state park and early restoration work on the Lindbergh house captured national attention; in the early 1930s, the flier's boyhood home became not just a Minnesota or even a Midwest attraction but an American one. According to the *New York Times*, "The reconstruction of the Lindbergh homesite holds an interest that extends far beyond the borders of the State." A Little Falls newspaper reported that "practically every daily paper in the United States" carried news of the park's establishment, and interest in the place rivaled that of the summer of 1927. Despite the ongoing construction work, as many as five hundred tourists visited the Lindbergh site daily. As to the genesis of the park project, papers gave credit where it was due: describing the site as a "neighborhood project," the *New York Times* recognized the citizens of Little Falls as first conceiving the idea to convert the private property into a public state park and maintaining interest to see it through the state legislature.[47]

Taking advantage of the spotlight, Little Falls readied for the park's dedication to be held on May 21, 1932, the fifth anniversary of Lindbergh's landing in Paris, with the hope that Charles and Anne would attend. However, the tragic kidnapping of the couple's first son on March 1 from their home near Hopewell, New Jersey, precluded any joyous celebration that spring. Charles Augustus Lindbergh Jr., dubbed "the Eaglet," was found dead a few miles from the family's property several weeks after he went missing. The kidnapping of the Lindbergh baby, called the "crime of the century," and the subsequent 1936 trial of Bruno Richard Hauptmann became the biggest news of the 1930s, surpassing stories on the stock market crash, New Deal measures, and the impending European war in readership numbers.[48] The kidnapping and trial greatly impacted the cultural milieu of 1930s America as the gruesome events "tapped into the deepest insecurities of the depression generation," argued a Lindbergh biographer. If the sanctity of home could be destroyed for the Lindberghs, a beloved and protected couple, then any average family could be in danger.[49]

Dixon Wecter, describing America's fascination with Lindbergh in the early 1940s, held that his name "implied mother and home and fundamental decency."[50] That the Lindberghs' first child was kidnapped from their own house, a place of safety and comfort, encouraged further sentimentalization of Charles's boyhood home in Minnesota. The impetus to restore and protect the Little Falls property likely strengthened as a result of the heartbreaking kidnapping that kept the Lindbergh family prominently in the news. Yet, in light of the tragedy, the dedication of the Lindbergh State Park was "indefinitely postponed." As the *Little Falls Daily Transcript* forlornly pronounced, "whereas the beautiful home and grounds have always prompted happy thoughts," following the family's loss "a tinge of tragedy must ever be associated with them."[51] In December 1935, after three years of enduring hysteria surrounding the case, the Lindberghs moved to England to escape further scrutiny.[52]

The Lindbergh State Park remained an extremely popular tourist destination even after the Lindbergh baby's abduction. In 1934, it welcomed approximately 55,000 visitors.[53] The next summer, before the Lindberghs left America for England, Charles and Anne made a surprise trip to Little Falls. The occasion marked Anne's first visit to her husband's hometown and Charles's first since his tour following his 1927 flight. The visit was informal, with the famous couple spending a good part of their time with park superintendent Engstrom at his combined confectionery and hardware store, a place Charles frequented as a child. Despite their mutual aversion to attention, the Lindberghs "made no attempt to dodge the public" on this trip. After visiting his old home, Charles "expressed both surprise and pleasure" at the changes made with the limited state appropriations. He must have enjoyed his time spent in his hometown for he returned at the end of August, unbeknownst to all that it would be his last visit for many years.[54]

Phase I: The Lindbergh Homestead Is Restored

The Lindberghs' trip to Little Falls in 1935 coincided with the second reorganization of the Minnesota state park system. The previous year, the National Park Service had appointed Harold W. Lathrop, an apprentice to the Minneapolis City Park superintendent, as supervisor of Civilian Conservation Corps work in state parks. This appointment sparked an examination of state park administration, which had been managed by the Forestry Division since the first reorganization in 1931. State legislators wanted to place the parks under the jurisdiction of someone trained specifically in park management rather than forest management. In July 1935, the Minnesota legislature created the State Park Division within the Department of Conservation, appointing Lathrop director. In his new role, Lathrop oversaw thirty state parks and coordinated CCC and WPA work within the park system.[55] This year also marked a shift in CCC policy, as the New Deal agency broadened its focus on environmental work in forests and farms from combating timber famines, reforestation, and fire prevention to developing recreational facilities to promote tourism in national and state parklands. As enrollees of the CCC and CCC-Indian Division, which had a regional office in Minneapolis, built visitor centers, park lodges, hiking trails, and campgrounds, they endorsed the message that playing in parks was good for Americans' public health. This ideology influenced the development of state park systems across the nation, including in Minnesota.[56]

After the federal government established the WPA in 1935, the Department of Conservation, official sponsor of all New Deal projects in state parks, applied for funds from the new work relief agency to finance improvement projects in eighteen smaller parks, one of which was the Lindbergh State Park.[57] The department submitted project applications to the state WPA office in Saint Paul, headed by the newly named state administrator Victor Christgau, a politician with a farming background. A former Republican state senator and U.S. congressman, Christgau had served as national assistant administrator of the Agricultural Adjustment Administration since 1933 but returned to Minnesota to take up the role of WPA state administrator. Christgau's three years in that position can be described as turbulent, as he faced direct challenges from Farmer-Labor politicians who demanded his removal.[58] However controversial a New Deal political figure, Christgau

consistently pushed for increased relief for farmers and supported the continued development of the state park system as a way to promote public health
and increase visitation to the state.

The Department of Conservation's initial project application for the Lindbergh State Park requested WPA funds to continue restoring the house, plant
four thousand trees and shrubs, construct a picnic area, develop two miles of
foot trails, and build a bridge over Pike Creek, a tributary of the Mississippi
River that ran through the property. In September 1935, Christgau approved
$26,204 for the Lindbergh home and grounds project. Of the total, $2,504
was earmarked for the restoration of Lindbergh's boyhood home and $23,700
for the improvement of the park grounds. When announcing the proposed
allocation of funds, the *Little Falls Daily Transcript* declared the project to be
"riding the crest of popular enthusiasm," and the *Salt Lake Telegram* of Salt
Lake City, Utah, called it "an excellent idea. . . . May the preservation of his
home keep his memory green!"[59] The visit of a moving picture cameraman
sent by the federal government to take photographs of the Lindbergh State
Park, as well as the flier's high school and nearby Camp Ripley, a National
Guard training camp where Lindbergh often had landed his plane, further
proved the mounting interest in the proposal.[60]

Although Christgau pledged his support for the project, final approval of
the allocation of funds rested in the WPA's central office in Washington, DC,
as was the case with all WPA projects. The central office delayed funding for all
Minnesota state park projects, however, until the spring of 1936 because state
applications were third in priority behind local and county projects, and Minnesota funds had been allocated elsewhere first.[61] Finally, in May, the ninth
anniversary of Lindbergh's historic flight, the DC office approved the Charles
A. Lindbergh State Park as project no. 2–573 with sponsorship of the State
Park Commission of the Department of Conservation and an initial allocation of $23,777, almost identical to the amount the state office had approved the
previous fall.[62] Work soon began at the project site, and a crew of forty to fifty
local WPA workers labored at the park during the summer of 1936.[63]

As the relief work progressed, the imperative of the WPA project remained
twofold: restore the house and develop the state park. Work first began on the
house, which was only minorly repaired in the early 1930s. The WPA project sought to finish returning the home to its appearance when Lindbergh
lived there as a boy, and the aviator himself made achieving that objective
possible by providing information in a series of letters penned to Dr. Grace

Nute, curator of manuscripts at the Minnesota Historical Society (MHS), between 1936 and 1939.[64] While curator, Nute gathered research on the Lindbergh family for a book she was writing about August Lindbergh, the famous flier's grandfather, who immigrated to the United States from Sweden. She communicated frequently with his grandson about the family's early years in Minnesota, and while doing so kept Charles updated on developments at the state park.[65] The celebrity's concern for the park's management demonstrated an enduring personal interest in what was to become of the property where he experienced much happiness and the freedoms associated with youth.[66]

Despite living in England during the years of the restoration, Lindbergh rarely failed to respond to Nute's questions about the old homestead where his family spent all of their summers between 1907 and 1920, as well as the three winters between 1917 and 1920 when he was in his late teens. Lindbergh's reminiscences of a childhood on the Mississippi River figured prominently in these letters. He described receiving his first gun and going on hunting expeditions with his father, playing on the farm with his beloved childhood dogs, sleeping in the screened porch overlooking the river regardless of the temperature, and swimming as a boy in the river and Pike Creek, named after explorer Zebulon Pike, who surveyed the upper Mississippi in 1805.[67] Lindbergh also explained to Nute the backstory of the mysterious "Moo Pond" located in front of the house, which, he wrote, "has given rise to a great many amusing stories." The Moo Pond was a small cement pool Lindbergh built as a boy in the corner of the ducklings' enclosure, named so because he was told as a child that "Moo" was the Native Chippewa (Ojibwe) word for dirt. He inscribed "Lindholm" in the concrete of the pool—a name his mother and father intended to call the farm, although "Camp" became the term generally used—and the names of his favorite dogs, Dingo and Wahgoosh, meaning fox.[68] His propensity for incorporating the Native Chippewa language into his childhood projects matched the popular image of Lindbergh as a spirited youngster fascinated by local history and enthralling stories of conflict between European settlers and Indigenous Americans.

In his letters Lindbergh also included detailed descriptions of the house in which he grew up, which helped determine the accuracy of the exterior and interior work undertaken by the WPA. For example, WPA workers added front stairs and foot railings to the home's side kitchen entrance and front porch (fig. 14), which did not exist when the Lindbergh family lived in the house. In 1936, upon receiving photographs of the property, Lindbergh

FIGURE 14: WPA workers repair windows at the Charles A. Lindbergh House, 1936. Source: Minnesota Historical Society.

wrote to the MHS superintendent to confirm that the exterior appeared the same except for the addition of those porch steps. The interior also presented some unfamiliar sights. The WPA added a new dining room table, floor rugs, and a dresser, but Lindbergh verified that the armchair and couch belonged to his family.[69]

A couple of years later, in December 1938, a representative from the Department of Conservation visited Nute at the MHS building in Saint Paint with a list of questions tourists asked at the state park. Nute, in turn, queried Lindbergh about the exterior and interiors colors of the house, the use of each room, and the remains of an old car beneath the porch. Lindbergh responded in detail: the house was painted white with gray trim; the roof was originally covered with red cloth and later slate gray material; the stone foundation was unpainted; the interior walls were of varying shades of brown and gray or unpainted; and the old automobile was probably his Saxon "Six," which his father bought in 1915 but souvenir hunters had dismantled in the aftermath of his transatlantic flight.[70] As requested, Lindbergh sent a hand-drawn diagram

FIGURE 15: Interior view of the living room with Charles Lindbergh Sr.'s restored bookcases. Source: Historic American Buildings Survey MN-79, Library of Congress Prints and Photographs Division.

identifying the rooms of the house, and in another letter he enclosed a rough sketch of the location of buildings on the grounds during the time he managed the farm between 1918 and 1920 while a student at the University of Wisconsin–Madison. The sketch included the main house, tenant house, ice house, chicken house, hog house, barn, and other secondary farm structures.[71]

Evangeline Lindbergh, Charles's mother, also became involved in the restoration of her old home, writing a long letter to her son during the WPA project. She fondly described the old "homeland," the summer residence where the family was "always content to be."[72] She agreed to send some of the family's original furniture from her home in Detroit to the park at a later date, if desired.[73] Meanwhile, Lindbergh's half-sister Eva Lindbergh Christie visited the home with Alma Kerr, the state director of the WPA Division of Women's and Professional Projects, to determine the placement of reproduction furniture. An additional $800 WPA grant in 1936 funded the manufacture of some of the pieces souvenir hunters had destroyed or stolen in 1927, including Congressman Lindbergh's bookcase that held his law books (fig. 15).[74] While Charles Lindbergh supported the restoration project overall, he expressed concern that the introduction of furniture at the historic site would require guards to protect it, which he feared would create an additional responsibility for the people of Little Falls. He was "particularly anxious" that the park not become a "burden" to residents, writing that he "did not want its upkeep to grow so complicated that it may become an obligation rather than an asset." When MHS superintendent Theodore Blegen wrote Lindbergh in October 1936 after a trip to Little Falls, he reassured Lindbergh that a WPA assistant helped guard the house, thereby relieving residents of the duty of neighborhood watch.[75]

Phase II: Developing the State Park's Recreational Areas

While Lindbergh family members personally contributed to the restoration of the house by providing objects and sharing recollections, the development of their former farmland as a recreational area proceeded with much less of their input.[76] The conceptual idea of the park was a modernized "frontier," a semblance of the environment the state's first European pioneers encountered when they began to settle the area in the eighteenth century. An idealized pioneer landscape visually complemented the popular view of Lindbergh as the ideal figure of frontier mythology and honored not just the famous flier but his family's ancestors as well. It prioritized the history of white settlers and ignored Native peoples, except for the role they played as a vanquished adversary in manifest destiny and colonization narratives. Thus, the project projected an image of benevolent nature and an inspiring past, where struggles between Native peoples and European American settlers were devoid of any violent or enduring hardships.

To help convey in physical form the intended atmosphere of the frontier age, the Department of Conservation embraced an architectural design called Rustic style for park structures at the Lindbergh State Park. Inspired by the work of mid-nineteenth-century landscape architects Andrew Jackson Downing and Frederick Law Olmsted, Rustic-style architecture first appeared in America's built environment in the Gilded Age summer camps of the Adirondack Mountains in New York and in designs at national parks, including the Old Faithful Inn at Yellowstone National Park and Le Conte Memorial Lodge in Yosemite Valley. Its practitioners advocated the graceful blending of natural and humanmade elements and the use of traditional craftsmanship in their designs.[77] The style extolled using native materials and made buildings accessories to nature rather than the principal features of a landscape.[78] Designers' emphasis on incorporating local materials and adopting traditional building techniques mirrored the cultural work of artists operating in other New Deal programs that highlighted Indigenous and regional traditions. For example, the Indian Arts and Crafts Board, part of Roosevelt's so-called Indian New Deal, also preserved and promoted traditional crafts and created economic opportunities in tourist markets for Native craftspeople.[79]

Rustic style, sometimes called "Government Rustic," became the predominant style of New Deal environmental building projects, especially those undertaken by the NPS, CCC, CCC-Indian Division, and WPA. In 1935, the NPS commissioned Albert H. Good, architect for the agency's State Park Division, to produce a pattern book of appropriate designs. In *Park Structures and Facilities*, Good articulated the guiding design principles of official NPS Rustic-style architecture: "Successfully handled, it is a style which, through the use of native materials in proper scale, and through the avoidance of rigid, straight lines, and over sophistication, gives the feeling of having been executed by pioneer craftsmen with limited hand tools. It thus achieves sympathy with the natural surroundings, and with the past."[80] Built structures were to harmonize with the natural setting and appear handcrafted, in effect fooling visitors about their period of construction to think they were relics from a bygone era. In Minnesota, architect and Saint Paul native Mary Elizabeth Jane Colter began to design park buildings using locally quarried stone and adobe, which garnered the positive attention of the NPS.[81] In the northern part of Minnesota where timber resources were abundant, wood was the most appropriate building material; in south and northwest Minnesota, stone; and in central Minnesota, the region of the Lindbergh State Park, a combination of stone and wood.[82]

Three entities jointly oversaw state park design and the execution of Rustic-style philosophy: the Minnesota Central Design Office in Saint Paul, a branch office of the NPS Regional Office in Omaha, and the Design Office of the State Park Division within the Department of Conservation. The Central Design Office, led by chief architect and Duluth native Edward W. Barber, typically designed park structures for CCC camps, while the Design Office of the State Park Division generally designed park buildings for smaller-scaled WPA projects, like the Lindbergh site, and sometimes duplicated or adapted designs coming out of the Central Design Office.[83] This streamlined practice of Rustic-style park construction created a level of cohesion among the projects of the CCC and WPA and an overall recognizable aesthetics to the state park system in general. Moreover, all design work and construction were subject to approval by the CCC or WPA park superintendent, inspectors from the NPS, the director of state parks, the NPS Regional Office in Omaha, and the State Park Division within the Department of the Interior in Washington, DC. This necessary coordination contributed to consistency in function and design. The NPS praised the style and craftsmanship of the work completed in Minnesota's state parks, and the publication *Park Structures and Facilities* featured the shelter pavilion at Scenic State Park and both the well shelter and Old Timer's Cabin at Itasca State Park, all in Minnesota, as excellent national examples of Rustic-style architecture.[84]

Mark Buckman, named superintendent of the WPA project at the Lindbergh State Park, oversaw the Rustic-style design plan that echoed the architectural simplicity of the Lindbergh home and respected the local terrain. Most structures were constructed between 1937 and 1939, during the second stage of the WPA project (the first phase began in 1935 with the restoration of the house). WPA workers built a park shelter, water tower, two bridges, a restroom building, stone water fountains, three parking areas, and three miles of foot trails. Draftsman Henry Nielsen and field engineer Lehmann Taylor of the Design Office of the State Park Division designed the park shelter, or "kitchen shelter" (fig. 16). A beautiful expression of Minnesota Rustic style, the T-shaped shelter of log construction sat on a concrete foundation that was covered with stone. WPA workers assembled the shelter's saddle-notched corners and exterior doors without the use of nails, which lent to its rustic appearance.[85] They painted the exterior of "prefabricated, well seasoned cedar logs" with a creosote and linseed oil stain to achieve a soft brown coloring.[86]

The park's nearby restroom building was of the same stained peeled-log construction and notched corners as the park shelter. One of the last WPA structures built in the park was the water tower. Completed in 1939 and constructed of native limestone, the three-story tower held five thousand gallons of water that pumped into the caretaker's residence (the historic tenant house), the restrooms, and the drinking fountains.[87] In addition to this construction work, which built up the tourist infrastructure of the site, WPA laborers created three miles of new foot trails along which they placed wood benches made using hand tools.[88] An additional grant of $16,000 in 1938— $12,000 from the federal government and $4,000 from the state—provided for

FIGURE 16: Rustic-style park shelter at the Lindbergh State Park, 2018. Source: Author photograph.

the construction of a game warden's building and a sewage disposal system. Once these two projects were completed, the Minnesota state WPA office no longer considered the Lindbergh State Park "an 'open pool' for labor." However, a couple of years later, in the spring of 1941, a final small WPA grant of $5,359 funded the reconstruction of the custodian's cabin and some grading and landscaping work, thereby completing the landscaping and construction at the Lindbergh property.[89]

While Rustic-style philosophy guided the aesthetics and architecture of the Lindbergh State Park, a conservation agenda—a key tenet of New Deal programs in middle American states—directed the development of the park's landscape features. Besides creating recreational spaces to boost tourism, the project sought to address the main ecological problem plaguing the park: soil erosion. The issue presented the biggest environmental challenge across the state, as approximately one-fourth of an inch of surface soil on Minnesota's farmlands disappeared yearly by 1936.[90] Little Falls newspapers closely followed discussions in Washington, DC, over legislation that would improve the state's farming situation, with the *Little Falls Herald* calling the Soil Conservation and Domestic Allotment Act of 1936 "perhaps the greatest movement for national preservation and progress ever undertaken by the government of our nation."[91] At the Lindbergh site, the banks along Pike Creek were badly eroded and the creek had started subsuming several large white pines, predicaments made worse by visitors climbing up and down the banks. To remedy the problem, WPA workers built a long stone retaining wall resembling riprapping and back-filled it with clay soil. Terraces, later to be planted with trees and shrubs with extensive root systems, additionally combated future soil erosion.[92]

The environmental activities at the Lindbergh State Park indicate that New Deal conservation policy influenced state park development strategies. The Lindbergh project, in its focus on treating the soil and reintroducing native species, supported agricultural reform like other New Deal programs, including the Resettlement Administration (later the Farm Security Administration) and the CCC.[93] Moreover, in both its Rustic-style design and conservation focus, the Lindbergh State Park functioned similarly to a CCC project; this likeness probably contributed to the positive public perception of the WPA project because Minnesotans generally regarded the CCC favorably.[94] A November 1937 article in the *Minnesota Conservationist* magazine attested to the success of the park's conservation strategy from both a romantic and practical perspective. The author, Theodore F. Meltzer, sentimentalized the historic site and applauded the WPA's efforts at environmental redress. He wrote, "High, heavily wooded banks along [Pike] creek and a dense forest throughout the rest of the land made the Lindbergh homestead a wilderness spot of rare beauty." He described the "quiet and bucolic stream flowing through a peaceful landscape," and "an old barn made of hand-hewn timbers, the relic of some early settler." Less romantically, though equally important, Meltzer praised the reforestation work that protected and replaced the impressive number of tree species native to the Lindbergh land.[95]

Besides addressing pressing environmental concerns and creating an attractive tourist site, the Lindbergh State Park project was important for pragmatic reasons: it offered work and a paycheck to citizens of Little Falls. Clarence Tuller, born in 1909, was one local worker who benefited from the federal relief project. Tuller moved to Lindbergh's hometown in 1920 with his father, Arthur Robinson Tuller, a photographer who captured snapshots of WPA activity in town. Clarence helped build the kitchen shelter at the

Lindbergh State Park and found employment as an unskilled worker at other WPA projects in Little Falls, including Camp Ripley, Pine Grove Park, and the water purification plant.[96] He remembered working with a crew of four others on the Lindbergh property, naming Ernest Como as his team's project leader. The supervisors were local men, although Tuller recollected officials, probably from the state WPA office in Saint Paul, visiting to inspect the work. When they came, the WPA workers paid them little mind since the officials "trusted people they put in charge of the project" to properly supervise on-the-ground work. More than half a century later, Tuller spoke of his time with the WPA fondly, recalling that "somebody [was] laughing and kidding with somebody all the time."[97] David Benson suggests that WPA workers and CCC enrollees employed in Minnesota's state parks during the 1930s were conscious of the ideological role their work played in "the Big Picture" of the New Deal. Through visits from officials, like those Tuller mentioned, and reading works-in-progress reports published by federal agencies, laborers connected their local work to larger processes of national recovery.[98]

The Historic Shrine and the Pioneer Past

In February 1937, Minnesota state administrator Victor Christgau received a letter from the WPA's Division of Information in Washington, DC, asking for names of projects that could be included in the historic shrine restoration program. Surprisingly, Christgau replied, "We have no projects which can be properly included insofar as the WPA is concerned unless you want Lindbergh Park near Little Falls where the State of Minnesota has taken steps to preserve the birthplace of the famous flyer."[99] Christgau also mentioned the restoration of Fort Ridgely, an 1853 U.S. Army fort associated with the history of the U.S.-Dakota War of 1862, and the Henry H. Sibley House, the home of the state's first governor (although Christgau probably meant the adjacent Faribault House, home of the early fur trader Jean B. Faribault).[100] Neither, however, were WPA projects at the time, yet there were other agency-sponsored preservation projects unfolding in the state not mentioned by Christgau that could have fit the bill, including the reconstruction of the Chippewa Lac qui Parle Mission at Lac qui Parle State Park and the restoration of the Longfellow House in Minneapolis.[101] It is unclear why the state administrator did not point to any of these restorations taking place, but his mention of the Lindbergh State Park suggests that to the state WPA office it was a project of greater notoriety than the others.

While Christgau's wording reflects some reservations about the Lindbergh project aligning with the WPA's historic shrine restoration program, Julius Stone, the associate director of the WPA's Division of Information, assured him that the agency "would like very much to consider Lindbergh Park as a project which should be included in the historic shrine material." Indeed, Christgau later wrote that the "WPA has been interested only in the Lindbergh Home," despite evidence of other restorations in the state later sent to him by Ralph D. Brown, state director of the Historical Records Survey. One of the projects Brown mentioned was the North West Company Fur Post at Grand Portage State Park, a CCC-Indian Division project developed jointly by the Consolidated Chippewa Agency and the MHS to reconstruct the old fur post stockade where Chippewa members and white fur companies traded. Although not a WPA project and therefore not eligible for inclusion in the historic shrine program, Brown's attention to this project, which he directed, recognized the important restoration and archaeological work of Native groups that contributed to the development of the state park system and historical tourism.[102]

When Stone requested a narrative description and photographs of completed or in-progress work at the Lindbergh site, the Minnesota WPA team complied and sent a summary of activities, twelve pictures of the restored house and park, and a special progress report written by Harold W. Lathrop, director of the State Park Division. Lathrop noted that although the state park was originally designated in honor of Congressman Lindbergh, the WPA rehabilitation "restablish[ed] the property as a tribute to the flier," the famous son's reputation eclipsing that of his father.[103] Lathrop also described in detail one of the highlights of the WPA work at the park: the reconstruction of a simple footbridge spanning Pike Creek. At the age of twelve, Lindbergh built this suspension bridge of barbed wire and wood slats so that he could avoid fording the stream when driving the cows to the barn to be milked. After hearing the story, project superintendent Mark Buckman searched for remains and found the rotted timbers that had supported the bridge along with rusted barbed wire from which the footbridge had been suspended. From these remnants, WPA workers were able to rebuild the bridge in its original location. Lathrop claimed that when looking at the reconstructed bridge, "one inevitably thought of the trail blazed across the Atlantic by the young builder."[104] It is clear that Lathrop hoped visitors to the park would make the connection between the built environment of central Minnesota and the brave, history-making adventures of Charles Lindbergh Jr.

Literature published by the WPA similarly depicted the Lindbergh State Park as a confluence of history and nature, which fit the larger image the federal agency crafted about the state in travel guides. For example, the Minnesota Writers' Project's "Minnesota Recreation Guide" described the state as a "Vacation Land." Its lakes, forests, and brushlands attracted water enthusiasts, anglers, and hunters, and made the North Star State a "modern playground, rich in the history of a not too distant past. The tradition of the explorer, the trapper and the hunter has not been broken."[105] The Mille Lacs Indian Trading Post near Onamia, Minnesota, about forty miles east of Little Falls, is an example of a business endeavor that capitalized on both history and nature, in addition to Native labor and talent, for tourism purposes. Members of the Mille Lacs Band of Ojibwe worked in the CCC-Indian Division and restored archaeological ruins and monuments both on Indian lands and in national and state parks. The efforts of enrollees helped to create attractions that positioned Minnesota as a vacation destination.[106]

Dr. Mabel S. Ulrich, state director of the Minnesota Writers' Project, oversaw the publication of the state's contribution to the popular American Guide Series, *Minnesota: A State Guide*, in 1938. When she sent draft chapters of the guide to the Washington office for approval, she received critical feedback from its editors. Ulrich wrote that her office was "completely baffled by the tendency of all federal editors to regard us as inhabiting a region romantically different than any other in the country."[107] The Washington staff wanted Ulrich and her team of writers to highlight the state's folklore, like the myth of Paul Bunyan and tales of meetings between Indigenous Americans and settlers on the frontier. Ulrich considered featuring Minnesota's rich Scandinavian heritage, but, as she told national Federal Writers' Project director Henry Alsberg, it was hardly "uniquely Minnesota."[108] Ulrich seems to have won that particular battle since the final publication has no chapter on folklore, but the guide highlighted the state's European pioneer heritage in other ways.

The writer of the essay on Minnesota's architecture, for example, described its buildings as collectively expressing the toughness, no-nonsense attitude, and aesthetic of the state's homesteaders. When they arrived to the United States, northern European immigrants "exchange[d] the picturesqueness and

discomfort of their Old World stone cottages and thatched barns for a plen-
titude of lumber."[109] With this lumber they built simple log cabins, perhaps
the most recognizable symbol of midwestern homesteading, which became
a common "usable past" in Minnesota during the 1930s, argued historian
Karal Ann Marling. In 1935, the *Saint Paul Pioneer Press* and the radio station
WTCN jointly constructed a log cabin from which to do their reporting
on the fairgrounds of the Minnesota State Fair. "By the use of the cabin
symbol," Marling posited, the press and newscasters "were packaged as just
plain folks—honest, outdoorsy, Minnesotans made of the stuff of the hardy
pioneers." The next year, the WPA built a cabin for the Forestry Division of
the Minnesota Department of Conservation, followed by other versions for
county fairs across the state.[110] As a representation of pioneer history, the log
cabin stood for both past hardships and "nature in the raw—the fields and
forests as they were before the plow and the saw and the pickax changed the
face of Minnesota forever." The log cabin, in other words, served as a vehicle
to return to simpler times, an invitation to play pioneer rather than engage
with the complexities of settler colonialism and its consequences on contemporary Minnesotans. The Lindbergh State Park's unpretentious Rustic-style
structures referenced the rather practical and unornamented architecture of
the log cabin, tapping into its historical and cultural meanings.

In addition to this physical homage to white settlers of the state, Minnesota Writers' Project authors penned pieces that glorified the historical legacy
of the area's Scandinavian settlers, especially Charles's paternal grandfather,
August Lindbergh (born Ola Månsson), who arrived from Sweden in the
early 1860s, and his father, the congressman. Author Curtis Erickson likened
the "privations and hardships" of the frontier environment the Lindbergh
men experienced to that of the Pilgrims on the East Coast two centuries
before them, advancing a continuum of the colonization narrative that connected the Lindbergh Park ideologically to the Whitfield House restoration.
According to Erickson, because of the exciting and wild environment of his
youth, Congressman Lindbergh became "the living expression of the spirit
of agrarian class-consciousness and frontier revolt."[111] Lindbergh Sr.'s legacy
as a populist defender of "the rights of plain people," particularly Minnesota
farmers, carried significant relevance and utility during the Depression years.
His ardent support of farmers and hardy frontier background resonated with
Minnesotans troubled by the agricultural and economic slumps of the 1930s.
His famous son, nicknamed "the blond Viking of the air," was born into this

rich heritage of Swedish frontierspeople, political sympathy for agricultural work, and a deep and abiding appreciation for the natural world.

Charles Lindbergh's own humble, self-effacing nature and quiet involvement in the WPA project demonstrated reverence for his own—and by extension the state's—pioneering forebearers. In his autobiography, published in 1978, Lindbergh described his childhood as "one generation beyond the Minnesota frontier."[112] While images of Chippewa canoes along the Mississippi River and the fur trading expeditions of early trappers loomed large in Lindbergh's mind as a child, they, too, featured prominently in the way Minnesotans chose to conceive of their state's history in the 1930s: a place of natural beauty and adventure which was challenging in a way that built character, yet conquerable. While speaking at the Lindbergh State Park in 1981, Lindbergh's daughter Reeve perceptively recognized that her father's boyhood in Minnesota "made the American past more accessible" to him and taught him to value the natural world that he, like others, felt was disappearing from the American landscape.[113] In a similar manner, as a so-called shrine restored by the WPA, Lindbergh's boyhood environment offered a visceral experience to Americans during the Depression who felt disconnected from their native soil and wanted to restore the important link among land, history, and personal character development. But it is important to recognize that the historical Minnesota presented and interpreted at the Lindbergh site was a self-consciously selective representation of the past, one where white settlers triumphed, the land provided abundant offerings, and Native peoples simply did not exist, reinforcing the misperception that Native cultures disappeared along with the frontier, despite their active involvement in CCC projects building the modern tourism industry.

Charles Lindbergh, the person, and the Charles A. Lindbergh State Park, the place, represented a society at a crossroads, with one foot in an agricultural past and one in the modern frontier, a world increasingly industrial and urban. Indeed, the section in *Minnesota: A State Guide* on "Agriculture and Farm Life" conceded that the Minnesota of the 1930s in reality "retained little of its pioneer flavor" as New Deal agricultural and economic programs helped modernize the state. When Anne Morrow Lindbergh first visited her husband's hometown, she acknowledged the homogenizing effect of modernity: she wrote in her diary that "Little Falls is just like hundreds of small towns in the West: the brick buildings on main street, the drugstore, the nondescript hotel, the gas stations, the plate-glass-store-front windows.

Not one building stands out in my mind; not one different from another."[114] Anne came to realize, however, that the Lindbergh State Park had become for Americans a historic place set apart from modern life, industrial growth, and commercialism. Years later, she said that the Little Falls property represented "the love of nature, the beauty of the wilderness, the sense of freedom and adventure, of the rivers, the traditions, and inheritance of Minnesota pioneers, courage, independence, and a sense of the boundless future."[115] The WPA project, enshrining the house as an homage to the aviator and the land as a tribute to previous generations, wrested the Lindbergh State Park from time. While "Main Street" in America became modernized, the Lindbergh State Park evoked the frontier past in its cultivated historic built environment.

Back in May 1936, in the middle of the WPA project, New York newspaper columnist Ward Morehouse visited Little Falls during a cross-country tour to uncover what "the Lindbergh legend" meant to the famous flier's hometown. Morehouse's syndicated story, appearing in the *Minneapolis Tribune*, noted that Little Falls had "a shrine in the form of a silver-gray frame house," and thousands of visitors "have come to prowl it, to gape at it, to touch it."[116] Lindbergh's boyhood home, the architectural *pièce de résistance* of the state park in the eyes of most visitors, was to the flier himself "of very secondary importance."[117] The famous celebrity, who became a passionate conservationist in his later years, always believed his old home's greatest asset was its ability to provide an environment in which his fellow Minnesotans could convene with nature. He envisioned the state park foremost as a place of pleasure for the people of Little Falls to enjoy, "where families can go on Saturday and Sunday and where children can enjoy playing in the creek and river." During the summer of 1937, superintendent Martin Engstrom wrote to his friend Lindbergh of the park's success and popularity, especially with farming families from the area: "Most any evening one will find ten to fifteen of our local families down there with their picnic lunch."[118] As a farmer in his late fifties expressed in a public forum published in the *Little Falls Herald*, "a happy environment, and making a good living depends on how well we are going to co-operate with nature."[119]

Because of Lindbergh's fame, the site became more than a local or even state attraction for families; it earned status as a *national* shrine. People from all corners of the United States flocked to the park throughout the 1930s because of its connection with a living celebrity—an American legend in the making—and its newfound "historic" designation. In 1938, approximately 86,000 people signed their names in the visitor register. The following year, almost 40,000 tourists from every state in the nation and several foreign countries visited during the summer season alone.[120] The Lindbergh Boyhood Home and State Park became a place where Americans could adulate the state's greatest hero, honor the significant contributions of the area's Scandinavian pioneers, and reaffirm their relationship with the natural world through recreation.

The WPA historic shrine project benefited from fortunate timing, as just after the park's completion the Farmer-Labor Party's control of the state came to an end and public opinion of the famous flier rapidly declined. The Farmer-Labor Party had bridged rural and urban interest groups, but the rise of the Minnesota Farm Bureau weakened farmer-laborer cooperation and led to the election of moderately liberal Republican Harold Stassen as governor in 1938. Contributing to the decline of the Farmer-Labor coalition was former governor Elmer Benson's unwise politicization of the WPA and ousting of its state administrator, Republican Victor Christgau, whom Benson feared was angling to campaign against him for the governorship.[121] Around the same time the Farmer-Labor Party's influence waned, Lindbergh became embroiled in politics surrounding World War II and allegations of sympathizing with Nazi Germany, which marred his reputation. However, as Roger Butterfield wrote for *Life* magazine in August 1941, despite Lindbergh's ascension as the face of the America First Committee, he still held great appeal to the American public who gravitated toward "the magic of his legendary name, the appeal of his personality, the sincerity of which he comes before the microphone."[122]

While public opinion of Lindbergh shifted drastically with the continuation of the war, the Lindbergh State Park fared better than the man. In part, it was because the historic shrine represented not Lindbergh's adulthood but his formative childhood years on the family farm and the settlers' narrative of his pioneering ancestors. The Lindbergh State Park was not tainted by Lindbergh's own meteoric fall from fame, his noninterventionist stance on the war, nor his increasingly rocky relationship with President Franklin Roosevelt. Its Rustic-style park architecture, scenic views, and welcoming trails and campgrounds remained a preeminent attraction for folks pursuing recreational activities, perhaps intrigued by the tumultuous career of Charles Lindbergh.

CHAPTER 5

Maverick's Pan-American Vision

La Villita Historic Arts Village, San Antonio, Texas

Re-creating the Little Spanish Village:
For the Promotion of Understanding and Peace between the American Nations;
To preserve Spanish and Southwestern Culture:
To foster Arts and Crafts . . .
Above all, this project will be for human good, for letting people learn how to make a living, to have a way of life; for constitutional democracy, peace and freedom.

Villita Ordinance, October 12, 1939

In 1941, hundreds of soldiers, dancers, and spectators flocked to the grounds of La Villita, a historic arts village in San Antonio, Texas, to enjoy performances by the Works Progress Administration's Tipica Orchestra.[1] A monthly dance series cosponsored by the San Antonio Recreation Council and the city provided hospitality and entertainment to U.S. servicemen and residents at night, while during the day young laborers in the National Youth Administration made supplies for army camps in La Villita's metalcraft and woodworking shops. Just two years prior, newspapers described this area, now a hive of wartime activity, as a "vile slum," an overcrowded little neighborhood with rundown buildings and no running water.[2] But by the early 1940s, with the political support of Mayor Maury Maverick, architectural vision of O'Neil

Ford, and financial contributions and labor of the NYA, La Villita became a focal point of the city's burgeoning tourist industry and wartime efforts. From 1939 to 1942, the New Deal agency helped transform La Villita into a training center for arts and crafts, a political backdrop for events with visiting Latin American dignitaries, and a central location of leading civic and military commemorative events.

Yet, La Villita's origin story presents a much humbler image than its vibrant appearance in the 1940s would suggest. The "little village" began as a haphazardly grown settlement on the east side of the San Antonio River sometime in the 1700s on the site of a sixteenth-century Coahuiltecan village.[3] Spanish colonization and the founding of Catholic missions, including the Mission San Antonio de Valero (the Alamo), in the early eighteenth century led to the arrival of the first residents of what would become La Villita. Adding to the population of Franciscan fathers and Indigenous Americans were soldiers stationed at the nearby Presidio of San Fernando de Bexar to protect the missions. Located between the Alamo and the presidio, La Villita over time became home to a mix of soldiers, Native peoples, and Canary Islanders, who resided in modest adobe and caliche homes. As the new Villa de San Fernando, also on the city's east side, attracted aristocratic Spanish families, La Villita became an unpretentious community colloquially said to be "on the wrong side of the river." Between 1793 and 1824, the Spanish government and Franciscans secularized the missions and a cavalry troop of Mexican soldiers stationed itself at the Alamo, their presence contributing to the racially mixed and nonpatrician nature of the area.

The composition of La Villita changed, however, when the San Antonio River flooded in 1819 and swept away many residences on lower ground. This natural disaster caused a migration of Spanish families to the higher-elevated La Villita, marking its transition from a common to exclusive residential area. Then, during the conflict over Texas independence in the 1830s, Mexican general Martín Perfecto de Cos surrendered to the Texan forces following the Battle of San Jacinto in 1835 in his La Villita home. After the Mexican defeat, Tejano landowners moved to the west side of the San Pedro Creek, thereby vacating the central areas of the city, including La Villita, for newcomers to fill.[4]

So began another chapter in the little town's history: starting in the 1840s, recently arrived immigrants from Germany, Alsace, Poland, France, and Switzerland settled and introduced new European institutions and architectural styles to the formerly Hispanic area. As San Antonio's agricultural economy and manufacturing industries blossomed after the railroads came to town in 1877, residents began to move to new suburban areas (fig. 17). Soon after, the Mexican Revolution in 1910 led to a rise in Mexican immigration to the city, creating a "race problem" that increased in the decade. Tightened visa requirements and border control policies of the decade ostracized Mexicans as "illegal aliens," and the group faced the additional weight of Jim Crow segregation laws and practices that minimized their economic opportunities.[5] As a result of these demographic shifts, in the early twentieth century La Villita evolved from a population of ethnically diverse homeowners to mostly low-income Mexican and Mexican American tenants living in increasingly dilapidated rental properties.

The story of the next phase of La Villita's biography, the NYA's transformation of a derelict area into a booming cultural center, began with a defeat:

FIGURE 17: La Villita (to the left of St. John's Lutheran Church, center) when it was a middle-class residential neighborhood, circa 1876. Source: University of Texas at San Antonio Special Collections.

Maury Maverick's loss of his congressional seat in the 1938 election. The thirty-nine-year-old San Antonio native had started his political career at home, first serving in public office as the tax collector for Bexar County from 1929 to 1932 before bursting onto the national political stage in 1935 to take a Democratic seat in the U.S. House of Representatives' newly created 20th District in Texas. For his political showmanship and forthright support of New Deal programs, Maverick earned a reputation as "one of Congress's most aggressive representatives and an ally to progressives across the country."[6] He supported stronger relief policies and a minimum-wage law, and staunchly defended civil liberties, becoming the only southern congressman to vote for the Anti-Lynching Bill of 1937. While this vote marked him as an "ultraliberal on racial issues," his commitment to democratic liberalism and economic equality in reality had limitations when it came to race.[7] However, at home in San Antonio, Maverick's support of the Congress of Industrial Organizations and the pecan shellers' protest in 1938 to secure better wages for Mexican American workers helped craft his liberal image.[8]

In one of his least popular political moves in his native state, Maverick backed Roosevelt's controversial plan to pack the Supreme Court in early 1937, while Vice President and Texan John Nance Garner and most of the Texas delegation in Congress broke with the president over this issue.[9] This decision may have cost Maverick his chance of holding on to his congressional seat; in the following year's election, he narrowly lost by fewer than six hundred votes to the more conservative Democrat Paul J. Kilday, who had the support of the San Antonio political machine and attacked Maverick as a "radical communist."[10]

Ousted from the national scene, the poster child for New Deal liberalism redirected his political efforts to bettering San Antonio, the town his grandfather Samuel Maverick helped found. Maverick ran for mayor in 1939 on an antimachine "Fusion" ticket, modeled after Mayor Fiorello La Guardia's movement in New York, with the goal to break the local machine and roll out a reform agenda.[11] Maverick contested the sitting mayor and his longtime adversary, C. K. Quin, whose political activity contributed to Maverick's loss in the congressional race the previous year. Maverick's victory in the mayoral election can be attributed partially to the Mexican American voting bloc: he supported unionism; counted as his friend local labor organizer Emma Tenayuca, who had led the pecan shellers' strike the year prior; and forced the Census Bureau to classify Mexican Americans as "white," all measures that earned him support. Compared to Quin, who used violence, scheming with employers, and Red-baiting to oppress the Mexican population and defuse threats of strikes, Maverick emerged as a potential political ally.[12]

As a reformist mayor, Maverick worked to improve public health and sanitation, reduce police corruption, decrease prostitution and gambling, and utilize federal programs to improve the lives of San Antonians, especially laborers and Mexican Americans. He envisioned San Antonio as "a place which is modern and streamlined to the hilt; where business can make money and people get decent wages and working conditions." Although intent on modernizing the city, he also wanted "to see all that is beautiful and quaint and historical preserved."[13] In that vein, he announced a plan to clean up the neglected La Villita near downtown as an NYA project, with the city of San Antonio as cosponsor. During his two years as mayor, La Villita became the grandest, costliest, and most visible civic endeavor of his administration. In sponsoring this project, Maverick employed preservation to advance four goals of his reform agenda: (1) improve housing and sanitation, (2) train Mexican American youth in employable skills, (3) develop tourism as an important sector of San Antonio's economy, and (4) form strategic alliances with Latin America.

Along those lines, Maverick conceived of La Villita's restoration and development into a historical attraction as an opportunity both to position San Antonio's youth, in particular young Mexican Americans, for economic success and to promote and preserve the city's history and culture while "cleaning up" unsightly areas. In a booklet he prepared to send philanthropic foundations asking for additional funds for the restoration, he introduced the underlying questions prompting the project:

1. What must she [San Antonio] do to see that her youth is given its start in a modern industrial world; to see that their training fits them for opportunities they fail to receive, untrained?

2. What shall she [San Antonio] do to preserve her heritage of the past, handed down to present-day Texas from her pioneer founders?[14]

Besides preparing San Antonio's Hispanic youth for employment and safeguarding historic resources, Maverick imagined La Villita playing a crucial role in fostering goodwill with Latin American countries. Toward the end of the NYA enterprise he explained, "the entire project had been built around the theme of Western Hemisphere solidarity."[15] Indeed, when popular journalist Ernie Pyle visited San Antonio at the end of 1939, six months into Maverick's term, he reported that the city was "grabbing firmly onto the new trend in American thought toward a Western Hemisphere family—the Good Neighbor policy." San Antonio was the "gateway" between the United States and Mexico, and La Villita served as evidence that the city was "the logical site of this whole new blending of Latin and Anglo-Saxon."[16]

Maverick clearly understood historic preservation as a political instrument and drew on his experience in the field at the federal level when promoting La Villita to local boosters, businesspeople, Latin American officials, and historical and civic organizations. In 1935, fresh into his first congressional term, the eager young Texan had introduced the National Historic Sites Act of 1935 in the House of Representatives at the request of Secretary of the Interior Harold L. Ickes.[17] Maverick lamented the focus on Anglo-Saxon history and eastern sites in preservation endeavors and long advocated for recognition of San Antonio's historic places associated with the period of Spanish colonization.[18] Once mayor, he worked within an already established local preservation movement, bolstering, rather than initiating, an impulse to promote the city's Spanish heritage through architectural preservation. The support of the influential women of the San Antonio Conservation Society (SACS) helped Maverick actualize his political dream. La Villita joined other successful local preservation projects of the period, including the restoration of the San José Mission, which became a National Park Service site, and the WPA's "beautification" of the San Antonio River. Although La Villita, like these other projects, was conceived, designed, and promoted by white political and professional actors, the city's Mexican American youth as NYA laborers were responsible for the idea's successful execution; their contributions of labor and authentic claim of heritage legitimized the project.

La Villita Transformed from "Slum" to "Showplace"

The inspiration to restore La Villita, according to Maury Maverick, came to him during a walk in the neighborhood "one moonlight night in June of 1939, soon after becoming Mayor."[19] Upon discovering that the local public service company owned much of the ramshackle area, he spoke with colleagues on the city commission who supported his idea to purchase the property. From that fateful evening, La Villita became the mayor's "pet project" or, according to his wife, "his child."[20] La Villita was not an unknown historic resource within the San Antonio preservation community, although it was ignored by city officials until Maverick turned his attention to it. In 1935, the SACS had recommended the area be targeted for renovation because of its significant history as the site of Mexican general Martín Perfecto de Cos's surrender in 1835 (fig. 18). The SACS began meeting at the Villita Street Art Gallery a few years later, and a member even proposed a restoration modeled after Los Angeles's Olvera Street, which had been turned into a *mercado* to celebrate the city's Mexican heritage.[21] Despite Maverick claiming La Villita's rebirth came to him in a lightbulb moment during a romantic moonlit stroll, he would have been aware of the SACS's interest in the area, as his wife, Terrell, was a longtime member. Regardless of the idea's origin, Maverick quickly took up the cause of La Villita's restoration and placed the project at the center of his mayoral program.

Maverick's friendships with powerful men in Washington formed during his time in Congress helped him access federal support for his hometown

FIGURE 18: Cos House at 514 Paseo de la Villita, 1934. Source: Marvin Eickenroht, Historic American Buildings Survey TX-33-A-6, Library of Congress Prints and Photographs Division.

project. In July 1939, he visited DC to obtain a pledge of $100,000 from Aubrey Williams, national administrator of the NYA.[22] Maverick probably sought funds from the NYA rather than another federal work relief agency, like the WPA, for two reasons: he was personal friends with Williams, and the NYA focused on youth relief and skills development, which aligned with Maverick's goal to aid young Mexican Americans.[23] The mayor succeeded in getting a commitment from Williams, and official presidential approval for La Villita as a federal project came later that fall with initial funding of $53,174 and part-time employment for three hundred "needy youth."[24] The NYA announcement of October 2, 1939, included a general overview of the restoration: "When restored the buildings will house arts and crafts workshops, museum and archives, and a center for cultural activities. The area . . . will be representative of the period before Texas became an independent republic. Architecture will follow the original style of the buildings. Handicraft will consist chiefly of work similar to that done several centuries ago. Cultural activities will include many Native folk celebrations. In this manner some of the life of several centuries ago will be preserved within a modern city."[25] As the La Villita project expanded, the NYA ended up contributing the full $100,000 Maverick requested and employed over one thousand youth workers during its three years of operation.

But when the NYA greenlit the project, the city of San Antonio had not yet acquired the property rights to La Villita, then defined as the area bounded by Villita Street on the north, South Presa Street on the west, Nacional Street on the south, and Womble Alley (later renamed King Phillip V Street) on the east. With the exception of three houses, the San Antonio Public Service Company still owned the land and structures within the desired project area.[26] In order to start advancing federal funds, Williams "required something in writing" that the city had authorization to enter La Villita. Thus, the city arranged a three-month lease with the San Antonio Public Service Company during the summer of 1939 during which the city "would have the right to enter on the property, oust people in possession and commence the improvement of the area." Afterward, the city could acquire La Villita property through exchange of real estate or agreement. Arguments among attorneys for the city, the NYA, and the title company extended the lease period, but on March 30, 1940, the city legally acquired La Villita by exchanging equal cash payments with the company. Maverick called the deal "the best any City Administration has made in 50 years."[27]

The timing of the NYA's approval of La Villita and the city's negotiations to acquire its property was fortunate because Maverick came under considerable political fire during his first six months as mayor. First, he failed to support Vice President and Texan John Nance Garner's candidacy for president, which angered the Democratic Party in San Antonio. Second, in August 1939, Maverick refused to cancel a meeting in the municipal auditorium of the Texas Communist Party. Facing intense public pressure to rescind the permit, especially from the Catholic archbishop and veterans' groups, the mayor defended the right of communists who had paid the ten-dollar rental fee to meet in the space. On the scheduled night, between five and ten thousand furious protestors interrupted the meeting, damaged the auditorium, and injured sixteen people in attendance.[28] Not long after the auditorium fiasco, Maverick faced another public relations problem. In October, a grand jury indicted the mayor for allegedly buying the poll-tax receipts of twenty-six voters during his congressional election in January 1938. In the end, the jury found Maverick innocent, but these incidents set a bitter tone early in the mayor's term.[29]

Under fire on multiple fronts, Maverick used La Villita—and its expansive political agenda—as a "feel-good" issue to win over San Antonians. While already announced as a federal NYA project, the city council's adoption of the Villita Ordinance on October 12, 1939, marked the "official birth" of the restoration.[30] The city spent several weeks producing a twenty-one-page ordinance, a beautiful document typed on parchment replete with historical maps and other illustrations. Blanding Sloan, NYA Arts and Crafts Division supervisor in San Antonio, designed the book's special green suede cover depicting the official seal of La Villita, which NYA youth workers made as a linoleum block print. The city printed two thousand copies to be sent to President Roosevelt and his cabinet, every state governor, and philanthropic foundations.[31] Although the NYA paid for the cost of labor, the city of San Antonio as project sponsor had to cover nonlabor costs, like that of materials and use of facilities, so Maverick hoped to receive additional money from outside sources.

The ordinance began by introducing the primary goals of the La Villita endeavor: "Re-creating the Little Spanish Village: For the Promotion of Understanding and Peace between the American Nations; To preserve Spanish and Southwestern Culture: To foster Arts and Crafts." To accomplish that, the plan called for the restoration of seven houses built in the period 1722 to 1850, the creation of a plaza and interior street, construction of a "museum-library-forum-restaurant," and landscaping of native shrubs and flowers. The ordinance also put La Villita in the context of its surroundings, referencing the historic Cos House and Villita Art Gallery across the street, and the development of the one-thousand-seat River Bank Theater under construction as part of the WPA's San Antonio River beautification project. La Villita, the document described, would not be a "mere lifeless copy, a sterile, strangulated art form," nor "a dead museum for mincing scholars, but a place for the living, and those not yet born," with modern sanitation facilities, plumbing, and electricity. Importantly, the ordinance declared the purpose of La Villita to preserve the "culture and traditions" of "under-privileged youth"—referring to the city's young Hispanic population—by providing training and work experience. The production of ceramics, woodwork, weaving, leatherwork, metalwork, and fine prints would "open the way to financial independence of those who pre-apprentice and produce here," expanding economic opportunities to a population that through discriminatory practices largely had been limited to agricultural work and pecan shelling.[32]

To forward the goal of pan-Americanism, the ordinance announced the naming of features at La Villita—"pronounced Vee-Yee-Tah," it instructed—after prominent Latin American figures. Hidalgo Street, a new inner street between five of the old houses, was named for Father Miguel Hidalgo y Costilla, a Catholic priest and the "Washington" of Mexican independence. Juarez Plaza, to the west of the Cos House, commemorated Benito Juarez, the Indigenous nineteenth-century president of Mexico. Finally, the new museum complex was to be named the Bolivar Building ("pronounced Beau-lee'-vahr") for Simón Bolívar, the Venezuelan political and military leader who led revolutions against the Spanish Empire in South American countries. Later, two of the restored houses would carry names of famous South Americans: the San Martin House, after the military commander who led Argentina's independence movement, and the Caxias House, honoring the Brazilian patriot Duque de Caxias.[33] The naming indicated an unconcealed effort to foster friendship between the city and Latin republics, and San Antonians took notice. The *San Antonio Light* reported "recent objections to naming all features of the Villita project after foreigners," and whether facetiously or not, city commissioners considered amending names to Juarez-Lincoln Plaza, Jefferson-Bolivar Building, and Hidalgo-Washington Street. The change never occurred.[34]

Defining La Villita as a public utility, the ordinance described it as the duty of the city to acquire unpurchased parcels of the property "at reasonable prices and as soon as practicable," and through eminent domain if necessary. Over the next few months, the city bought the remaining privately owned lots within the boundaries of the La Villita area.[35] While the city maneuvered legal ownership, NYA workers began the process of removing people from the neighborhood, which was repeatedly referred to as a slum by those orchestrating the restoration. Maverick himself called La Villita a "squalid, miserable, decadent slum," and the commissioner of fire and police described it as "a hangout for winos, all sorts of . . . vice, and a terrible looking, dirty neighborhood."[36] An NYA staff member similarly pronounced La Villita to be "one of the worst slums in the State, where filth and disease prevailed and rats, mice, vermin, and people lived," and Texan newspapers used similar language in articles reporting the site's transformation.[37] Ernie Pyle echoed the politicians, planners, and media: "It was one of San Antonio's worst slums, and San Antonio is rather famous for its slums." In the end, though, he predicted La Villita would be "a minor Williamsburg" when completed, referring to the successful rebirth of Virginia's colonial capital.[38]

There is no doubt that La Villita was in an advanced state of disrepair at the time; while details of the neighborhood are not well represented in textual records, photographs of the area pre-restoration visually illustrate the low economic status of residents, lack of public services, and overcrowding (fig. 19). A visiting NPS historian from the Santa Fe office described its condition as preservation-through-neglect: "La Villita has been kept in a remarkable state of preservation, largely because for many years, during the rapid development of the city, it served as a slum area for the poor people of San Antonio."[39] Maverick also credited the tenants who lived there for the area's survival, writing that "the section has endured . . . largely due to the loyalty of its families, many of whom cling to their faithful old houses of stone and adobe."[40] In other words, La Villita was saved both because it had not been a desirable location during the commercial development boom of the 1920s *and* because residents forced by poverty to live there treasured their surroundings, however impoverished. This perception created a romanticized view of La Villita and the Mexican families who lived there as entirely removed from the modern age, a place whose history stopped before modern San Antonio emerged. It was a convenient portrait to paint at a time when oppressive segregation that

FIGURE 19: La Villita pre-restoration, 1939, before NYA workers cleared the site. In the foreground, a wooden sign for a "National Club" warned trespassers, "If you don't like this turn your head the other way." Source: University of Texas at San Antonio Special Collections.

mirrored the Jim Crow South accounted for substandard living conditions, not residents' choice.

Maverick's and others' use of the term "slum" to describe La Villita was not simply a derisive or idealized epithet; this rhetoric linked the restoration to the mayor's efforts to improve housing conditions in the city, especially for residents of Mexican descent. Early in his mayoral term, Maverick explained his political agenda for the city as "cleaning up and straightening out."[41] San Antonio had the worst housing conditions of the three biggest Texan cities (the other two were Dallas and Houston) and the largest Mexican population of any Texan city, which grew after tens of thousands of Mexicans moved north from south Texas and Mexico during the Mexican Revolution from 1910 to 1920. Throughout the Depression, even larger numbers of Mexican workers flocked to cities seeking relief, especially agricultural workers who were excluded from receiving aid by New Deal farm policy. Moreover, federal programs were administered at the local level, which meant that often hostile officials, employers, and unions perpetuated the repressive racial culture while executing their duties.[42] Political historian Mae Ngai has argued that Mexicans' situation in Texas was particularly dismal. Both Mexicans and Mexican Americans lived in segregated *colonías*.[43] By the end of the 1930s, most of San Antonio's 90,000 Mexicans and Mexican Americans, about 40 percent of the city's population of 250,000, lived in a four-square-mile barrio west of downtown because of discriminatory housing contracts.[44]

Within the area's one-mile core, more than 12,000 people lived in "decrepit wooden flimsy shacks often built by the occupants and also converted from their former use as horse stalls." These "corrals," as they were called, were typically no bigger than ten by ten feet and housed from one to four families.[45] In addition to small and cramped housing, Mexican neighborhoods lacked basic public amenities like lighted streets, adequate sewer systems, paved grounds, and green space, and they were rarely visited by city cleanup crews or serviced by police or fire departments. A lack of indoor plumbing meant that communal toilets or pit privies shared by fifteen to twenty families were common and a major source of disease.[46] As a result, the westside barrio became a "concentration of racialized poverty" by the 1930s, according to urban studies scholar Laura R. Barraclough.[47]

La Villita, while not part of this specific barrio, shared similar housing conditions. A key tenet of Maverick's broader political program was to decrease health problems related to tuberculosis and syphilis, particularly high among the Mexican population because of inadequate housing.[48] Prior to Maverick's election, the San Antonio Housing Authority had applied for funding to improve the city's conditions from the U.S. Housing Authority, a New Deal agency created under the Housing Act of 1937. The agency awarded federal funding to five projects. With segregation enforced, two were built specifically for Mexican Americans: Alazan and Apache Courts, located west and north of the Alazan and Apache Creeks in the west-central part of the city. While in Congress, Maverick had secured $4 million to complete the "slum clearance" housing project, which, he proclaimed, "should be one of the best methods of teaching the Mexican people sanitation."[49] La Villita, too, would be a "model for sanitation and slum clearance improvements," helping ameliorate the city's housing stock and rehabilitate a blighted area.[50] The Alazan and Apache Courts, however, rehoused only five thousand Tejanos, while most of the remaining fifty thousand Spanish-speaking residents continued to reside in low-quality houses and shanties.[51]

Little is known about the actual displacement of the people who made their home in La Villita. While the lease with the San Antonio Public Service Company gave the city the right to "oust people in possession," and the ordinance authorized the city to "physically clean" the area, there is no documentation for what happened to the 119 residents who lived there. What is known is that their removal in 1939 happened quickly, sometime between July 20, when the lease started, and August 3, when 110 NYA workers began "clearance, demolition and preliminary grading."[52] Figure 19 shows the poor physical condition of La Villita right before clearance began; discarded bed frames, remains of a car and other mechanical equipment, jerry-built wooden structures near old adobe and caliche houses, and chunks of broken building material cluttered the downtown area where nature still intruded. In the foreground, a wooden sign for a "National Club" warned trespassers, "If you don't like this turn your head the other way." One cannot say whether the sign appeared before the NYA cleanup crew arrived.

A "Progress Report" published in November 1939 by the NYA included an astounding stream-of-consciousness description of the process of removing tenants, the style intended most likely to capture the discomfort and disgust of those charged with the unseemly task:

> The first task of the removal of human wreckage was difficult and one requiring great delicacy—(as well as high immunity),—lest the city be guilty of causing unnecessary suffering among people already pressed beyond human semblance or endurance;—(and lest the re-movers be stricken with *what* disease) . . .—119 resident humans . . . seven houses—broken—half fallen—half gone—an area-way—"Hoover Village" reminiscence—dry-goods-box-scrap-tin—cell-size shanties with migrant strangers uncounted—"efficiency apartments"—space for a cot if you have one . . . three out-of-door toilets—the flushing type—all awaiting the plumber—five cold water taps—give this tubercular woman time—help her husband find new quarters—move that family of children—What? No beds—lodgers too—bring your own cot—five cents a "flop"—these two rooms take 26—another family—6 people—careful! . . . Can't stand the stench!—even the ground smells—what? The re-movers are getting lice and vermin—bring

the disinfectant—venerable old man—26 years here—26 years is 46 loads of cherished junk to be moved—that old derelict with a bundle of ragged clothes . . . ten crippled and diseased dogs—catch those mangy cats—call the Humane Society!

The report proclaimed that the "eviction-by convenience" was carried out with "every possible kindness and human consideration," along with a fleet of city trucks.[53] Beyond that, no newspaper covered the story in any detail; the people who lived in La Villita before the restoration were invisible in contemporary accounts. The commissioner of fire and police "assume[d] that they went into public housing," and it is possible that the residents applied for the insufficient spots at Alazan and Apache Courts, then under construction.[54]

Once the residents had been forced to relocate, work of preparing the project site began. NYA laborers leveled the ground and demolished shacks, lean-tos, sheds, outdoor toilets, an early twentieth-century coffee warehouse, and two brick commercial structures dating to around 1890, whose quarter million bricks were used later to build walkways and an onsite ceramics kiln.[55] By October 8, they had finished clearing the site of debris and any evidence of previous residents.[56] In total, workers removed from La Villita 25 truckloads of scrap lumber to be used for kindling in public parks and 162 truckloads of junk, the reminders of "broken lives that had been living in these ruins of ancient buildings."[57] By the next month, La Villita "lost its designation as a slum" and became "a San Antonio showplace," according to local newspapers.[58]

Two publications came out in the fall of 1939 that promoted the project both locally and nationally. The first was *Old Villita*, a booklet compiled by the Texas Writers' Project for the WPA's American Guide Series. The work drew on historical accounts from the past three hundred years to present a history of the "oldest remaining residential area of the city," from its roots as a Coahuiltecan village to its decline in the twentieth century.[59] Maverick sent the booklet to interested parties across the country and put it on sale at the Spanish Governor's Palace and the Alamo, two popular local historic attractions.[60] In November, the second propagandist publication came out: the aforementioned NYA progress report documenting the project's "aims and historic significance" and the city's political ambitions. The work included a short history and timeline of La Villita, images of maps and paintings from the eighteenth and nineteenth centuries, photographs of buildings pre-restoration, and architectural drawings of the existing adobe houses and the to-be-constructed Bolivar Building. To add intellectual weight to the assessment, the report included "authoritative opinions" from University of Texas at Austin scholars with expertise in Latin American and Texan history. According to the report, young workers employed by the NYA already had acquired practical trades skill while preparing the project site, such as how to safely demolish a building, lay foundations, record building measurements, mix mortar, and make adobe block. The city's youth soon would gain employable skills in arts and crafts once the training center was established, further broadening their job prospects to the construction and arts fields.

Maverick sent the progress report to all NYA state administrators, who praised the project, calling La Villita "a fine cultural stimulus for the youth of San Antonio" and "a guide which will show prospective sponsors what type of rehabilitation work NYA can perform, particularly in the restoration of historic landmarks." The state administrator of Ohio expressed envy for his counterpart in Texas, Jesse Kellam, for the opportunity to work on this "unusual enterprise," and the progress report became required reading for California's office and field staff.[61] Maverick's publicity efforts clearly worked to position his project simultaneously as a symbol of the NYA's high-quality historical work and as a demonstration of his skillful political leveraging of New Deal money to improve his city's tourist infrastructure.

Architect O'Neil Ford's "Indigenous" Village

In the summer of 1939, Maury Maverick called on his friend David R. Williams, an architect and the deputy administrator of the NYA who the mayor knew from his time at the University of Texas at Austin.[62] Williams was well acquainted with La Villita; in the 1920s, he had traveled around Texas with his young apprentice, O'Neil Ford, to document representatives of humble domestic structures. They visited La Villita as part of their survey and later worked together on the publication of two essays on "indigenous" Texas architecture.[63] Maverick knew of Williams's affection for La Villita and approached him about the NYA restoration project. Williams agreed to assist on one condition: that Maverick take on his protégé Ford, for he was with "whom I have discussed every detail of how a complete restoration of La Villita to its original beauty should be planned and could be achieved."[64] The mayor acquiesced, and Ford arrived in San Antonio in July to begin work on La Villita as consulting architect.[65]

Ford's overarching architectural vision for La Villita was to showcase the growth of Texas architecture between 1722 to 1850, when the area's "last radical changes, repairs, and redecoration occurred."[66] In other words, Ford chose to restore La Villita according to a period of significance rather than a specific date "to present a more varied example of systems and periods of time," which marked a departure from the Dock Street Theatre and Whitfield House projects. The result at the Texas site would be "no museum of buildings, and no replica, but a living evidence of how this indigenous architecture grew."[67] This decision meant removing any structures that postdated the mid-nineteenth century, including the aforementioned two Victorian brick commercial structures and twentieth-century coffee warehouse, and "all poorly built and extraneous additions,—gadgets, scroll work, smooth-pressed brick, cheap, mill-made doors, etc." adorning the houses that also appeared to postdate 1850.[68]

To Ford, La Villita exemplified "a direct and unaffected expression of our forefathers within the limitations of materials, climate, and the times," and "the absence of any affectation, personal conceit, 'prettiness' or self-glorification." What made La Villita special—indeed, authentic—to Ford was its simplicity and representation of a natural blending of styles corresponding to residents' diverse backgrounds. Ford's approach, which a biographer described as "vernacular authenticity," afforded the architect the flexibility to incorporate various native materials, including adobe, mud, caliche, half-timbers, and fieldstone, and different architectural elements drawn from the Spanish, Mexican, German, French, and Polish builders who resided in the village during the chosen period of significance.[69] Ford respected the integrity of the existing native earthen materials and had NYA laborers use traditional building methods to restore the extant dwellings. Besides addressing the historic structures, an integral part of creating La Villita was the landscaping; the specifications for La Villita, drawn by city architects with whom Ford consulted during his planning, included the lands, houses, gardens, wells, cisterns, outdoor patio, and Juarez Plaza. Maverick named as "city forester" Stewart King, who oversaw the planting of cacti, Spanish daggers, basket grasses, mountain laurel, manzanita, wild plum, and other native Texas fauna, along with a dozen live oaks on Juarez Plaza.[70]

The seven restored houses in the block bounded by Nacional, South Presa, Villita, and King Phillip V Streets—the core of La Villita—were generally simple, one-story rectangular structures between 1,000 and 1,500 square feet with hipped or gabled tin roofs, recessed entries, and wide front porches with shed roofs that extended the length of the facades (as seen in the house draw-ings in fig. 20). Despite the period of significance beginning in 1722, none of the extant houses were likely built during the eighteenth century, and it is estimated that most actually postdated 1850 (see table 1). Of the seven so-called restored houses in the main block of La Villita, the NYA weaving building was not a restoration at all but a reconstruction. Workers demolished the over two-hundred-year-old structure because "the foundation and walls were beyond repair," according to Fenner Roth, NYA district director. In its place a new structure was "built along the lines of the old one but guaranteed to fit the modern way of life," which meant accommodating a restroom.[71]

Three properties adjacent to the bounded area of La Villita were restored as part of the NYA project as well but not used for youth training purposes: the Cos House (constructed pre-1835), Villita Art Gallery (also called the Dashiell House, build in the mid-1800s), and the Puppet Theater (which

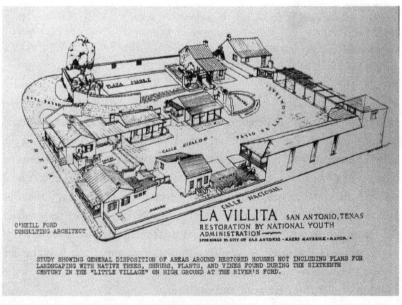

FIGURE 20: Architect O'Neil Ford's sketch of the restored La Villita. Source: "La Villita Progress Report," November 1939, Box 6, Folder "Texas Information Reports," Entry NC-35 86, RG 119, National Archives at College Park, Maryland.

TABLE I. La Villita buildings restored by the NYA

Building No.	Name(s)	NYA Use(s)	Construction Date	Material(s)
1	Guadalupe House	Ceramics	c. 1870	Caliche
2	Hesller House/ Canada House	Woodworking	c. 1870	Caliche
3	Caxias House/ Tejada House	Woodworking, Exhibits	c. 1855	Adobe
4	San Martin House/Herrera House	Societies Building	Pre-1854	Adobe
5	N/A	Weaving	Rebuilt 1940	Adobe
6	Bowen/Kirchner House	Dept of Information	c. 1860	Caliche
7	Losana House	Metal Crafts Etc.	c. 1860	Caliche

Source: Estimated dates are drawn from the document "History," n.d., Folder "La Villita History Gen'l, 1817–1940," SACS. Building materials are drawn from "La Villita Progress Report."

became the Little Church, constructed in 1876). The Cos House, where Mexican general Martín Perfecto de Cos signed the Articles of Capitulation in 1835, sat at 513 Villita Street opposite the northeast corner of the central core of the village. The Historic American Buildings Survey documented the house in 1934, and the SACS saved it from demolition and partially restored it before Maverick initiated La Villita as a work relief project. The SACS held club meetings there before transferring the property to the city.[72] The Cos House was the first building restored under the NYA; young workers retiled the floors, whitened the walls, refinished the ceilings, and repaired interior woodwork and the original fireplace. They also enlarged the building for "modern improvements, in the interest of public service," which included the addition of a kitchen, restroom, and patio.[73]

The grandest unit of the "restoration" project by far was the construction of the Bolivar Building, designed to include a branch of the Carnegie Library, the Witte Museum (of Texan history and culture), and a tourist information center. Because of the building's expected high costs, Maverick sought grants from nonfederal sources, including the Brookings Institution, Laura Spellman Foundation, and Rockefeller Foundation, but had no luck, despite the last's financial support of Williamsburg's restoration in Virginia.[74] However, in December 1939, the city announced that the Carnegie Corporation of New York City had awarded a $15,000 grant for the library-museum to supplement NYA funds. The federal agency furnished the plans, supervision, and workers, while the city paid for the materials using the Carnegie grant.[75] With prodding of the state NYA office by Maverick, who was "greatly humiliated and embarrassed" by the delay in construction after funds had been secured, work on the Bolivar Building began in the summer of 1940.[76]

Architect O'Neil Ford designed the Bolivar Building in the style of 1850s frontier military posts, like that of the nearby Indian Trading Post on Flores Street, "for the sake of authenticity of type within the period." Ford intended for the two-story, 120-by-30-foot stone building to be built of adobe and caliche blocks, which would "harmonize with the old buildings in color, texture, and general mass and line." In the end, however, perhaps because of material costs, the building received a limestone veneer.[77] The lower level included the library and museum, which would exhibit Hispanic collections, while stairways decorated with iron-grille work led to the upper story's large entertainment hall.[78] Altogether, the construction of the Bolivar Building called for doubling the size of the NYA workforce, bringing the total number of youths employed at La Villita at one time to 330, working sixty hours a month.[79]

Ford Contends with Maverick's "Sweetie-Pie" Spanish Vision

Ford's overall vision for La Villita as an unpretentious amalgamation of generally modest structures conflicted with the mayor's grandiose and enchanting portrait of the site. Maverick wanted La Villita to be a "showpiece," a tourist attraction and clear representative of Spanish influence on San Antonio. In other words, he wanted to see colorful floor tiles, wrought-iron detailing, and other identifiably "Spanish" architectural elements. According to Ford, Maverick imagined a "rather typical Spanish suburban idea . . . [a] sweetie-pie thing."[80] The mayor clearly explained his vision in the introduction he wrote for the Texas Writers' Project's publication *Old Villita*: "While *dulce* vendors squat in the shadow of the little courts and *tamale* women swathed in *rebozos* scent the air with their pungent pots of steaming edibles, strolling *caballeros* wearing broad, braided *sombreros* and short jackets of green silk will sing to their own stringed accompaniment of the songs of old Mexico and Spain—and the notes of the guitars, the odor of *masa* cooking, the soft voices of Latins, will help roll back the years to the time when these songs, these houses, these were San Antonio."[81] Maverick's ideas aligned with the general nostalgia and appreciation for Spanish heritage that had fueled the women-led preservation movement in San Antonio for decades, resulting in campaigns to support the preservation of both the Spanish Governor's Palace and the Alamo.

The ideological conflict between the enterprising Maverick and the lead architect quickly became apparent as Ford had to confront the popularity of the mayor's idealization of La Villita. About a year into the project, Ford sent NYA state administrator Jesse Kellam a report describing "the conflicting opinions and innumerable prejudices of the citizens of San Antonio" regarding La Villita. He found it "distressing to discover" that many residents, including members of the SACS, were disappointed by the restraint of the "mature architecture" that lacked the "claptrap" of brightly colored floor tiles, fat arches, Spanish barrel-tile roofs, or fancy iron grillwork, the very Spanish-style trimmings that Maverick dreamed of for the site. Ford broadly blamed his fellow architects, so-called professionals who had "hoodwinked" the public into believing they understood historic architecture by assembling a "hodge podge" of styles and adding "stuck-on appendages" to new designs. At La Villita, Ford maintained, "at no time do we expect to affect picturesque and 'sweetness' at the expense of good sense and structural honesty."[82]

Ford's report to the NYA reveals a stark incongruity between the extant built environment and the rhetoric surrounding the Spanish "little village." The physical structures of La Villita most directly represented the primarily German people who settled in the neighborhood beginning in the 1840s, not the area's earlier Spanish residents. While some San Antonians argued for more evidence of the German influence on La Villita in the restoration—the city building inspector and a city architect, for instance—the village continued to be touted primarily as a symbol of Spanish heritage.[83] San Antonians seemed eager to embrace a romanticized portrait of their city, which visualized the Spanish colonial past and deemphasized the more diverse, quotidian existence of later residents. The multicultural and more commercial history of La Villita in the mid- to late nineteenth century was well documented in the Texas Writers' Project's booklet but most newspapers framed the ongoing restoration according to the mythologized version advertised in official project rhetoric. For example, take the *Mexia [Texas] Weekly Herald*'s description of La Villita, whose narrative ends in the 1830s: "Through the centuries it has survived Indian raids, rebellions, epidemics, famines, wars, and the rise and fall of governments. Its homes have entertained Spanish viceroys, Franciscan padres, conquistadores, cavaliers, explorers and men prominent in the affairs of the Texas Republic."[84] The German, French, Alsatian, Swiss, Black, Mexican, and Mexican American residents are mentioned nowhere.

A strong proponent of restoring La Villita as a romanticized representation of the Spanish days of yore was Mary Rowena (Rena) Maverick Green, cofounder of the SACS and first cousin of the mayor. She wrote to the local newspaper with praise for the project but also encouragement to "hold carefully to what is typically old San Antonio." She offered as emblems sloping tin roofs either rusted or painted red, and evidence of the old ditch system with irrigated gardens. She even suggested reinstating representatives of the very group of residents displaced to undertake the project: "Here we should have some Mexican families living—for nothing could be as pleasing or charming. At some spot along the ditch a thatched shed . . . and a few tubs—and what an ideal place for women to gather to do their washing!"[85] Perhaps if Green had visited La Villita before the restoration she would have seen an image much like the one she envisioned, but she would not have thought it charming. Green's bold proposal to exhibit Mexican families for added charm to the site illustrates, first, the conflation of Mexicanos and Hispanos, and, second, the pivotal role Mexicans and Mexican Americans played ideologically in Spanish heritage projects while most Anglos in reality disassociated them from their contemporary lived experience. As Laura Hernández-Ehrisman has argued, "Mexicanos were both socially displaced and symbolically centralized" as Anglos fabricated their heritage through place-making and customs in San Antonio.[86] The local preservation community's interest in the "Spanish" past might have been genuine, but their curiosity did not typically extend to political activism or inclusion of cultural representatives in city projects that capitalized on broader Hispanic or Mexican heritage.

Not all San Antonians, however, ignored the more recent chapters of La Villita's history. One resident wrote to the *San Antonio Express* with information about "the substantial citizens" who had called La Villita home during the previous seventy-five years, with emphasis on the late nineteenth century.

The article, published in December 1939, included an incredible directory of residents' names, addresses, and occupations. The neighborhood counted a popular boardinghouse owner, banker, county physician, fortune teller, shoemaker, bookkeeper, confectioner, hardware merchant, dressmaker, grocer, and traveling salesman.[87] During this period, the majority of La Villita residents owned their homes, many of which were demolished by the 1930s, and carried last names announcing their German, French, and Polish roots. *Old Villita* documented these European immigrants in a section called "Newcomers to the Little Town," describing the jovial times of the artists and craftspeople in the village as they celebrated birthdays and Christmases and established community centers like the German Methodist Episcopal Church and German school. The architecture reflected its new German owners as they repaired their homes with plaster and mortar and added steep-pitched gable roofs that extended in the front to create covered porches.[88]

While a Spanish version of the "little village" remained central to Maverick's vision and use of the site, Ford ensured that the built environment in its modest styles and native materials expressed this broader history of settlement in the neighborhood. It is important to note, however, that all of the 1930s sources previously mentioned largely ignored the people who resided in La Villita in the four decades immediately preceding its restoration. Black residents, who began living there in the late nineteenth century, and Mexican and Mexican American tenants, who rented homes in the early twentieth, received minimal, if any, attention. Also overlooked were important local landmarks in the La Villita area, like St. Philip's College for Black girls, which operated between 1898 and 1917, the city's first African American Episcopal Church, and twentieth-century commercial structures that housed businesses like a bakery and auto repair shop.[89]

In the end, the restored La Villita became a combination of Maverick's idealized Spanish hamlet and Ford's model vernacular Texan village, but it did not happen seamlessly. David Williams had predicted tension between Ford and Maverick before the project began, telling the mayor, "There will be times when you become so exasperated with him [Ford] that . . . you will surely kick him off the job, for I know both your temperaments and I know you will clash."[90] Maverick did become frustrated with Ford but not primarily because of his architectural vision, however seemingly incompatible it was with Maverick's own. Rather, he resented the architect's inability to stay in San Antonio for long periods of time. Ford's position as consulting architect meant he received a salary of ten dollars a day plus expenses; the La Villita project was not a full-time job, and he accepted other work under his Dallas office of Ford & Swank that required him to travel around the state.[91] NYA officials as well as Maverick noted Ford's frequent absences. When the NYA advisor for arts and crafts visited La Villita in June 1940, he reported that supervisors "indicated the contacts with Mr. Ford are too limited, and his guidance is too casual. . . . Their dependence on Mr. Ford for specifications has limited their operations to his capacity to keep pace with their production possibilities."[92] Nonetheless, Ford's designs generally pleased Maverick and the NYA officials, and the mayor kept his word to David Williams; Ford remained involved in the project throughout its entirety despite disagreements.[93]

La Villita Becomes an NYA Training Center

The physical work of excavating, restoring, constructing, and landscaping the site was only phase one of the La Villita project. After the NYA laborers completed that work, La Villita entered phase two, establishing itself as a training center for San Antonio youth in six different arts and crafts shops: ceramics, weaving, design, metalwork, woodwork, and soft metal (copper and pewter).[94] Throughout the NYA training center's period of operation, over a thousand mostly Mexican American boys and girls between the ages of eighteen and twenty-four developed employable skills.[95] A new shift of workers was substituted every two weeks to create more job openings for youth from relief families, so there was constant labor turnover at the site. The NYA trainees learned to design and make ceramics, glass, tile, pottery, and sculpture. They gained carpentry skills through production of furniture, wood carvings, and puppets, and they were taught the art of textile fabrication, weaving, leatherwork, metalwork, and fine prints. The products they created were of professional quality, and most were sent to NYA resident centers or construction projects around the state. Some items were made especially for the mayor's office, such as handwrought iron replicas of an old Spanish key found on site during excavation, and other pieces were sent to the White House to showcase their quality, including a coffee pot engraved with President Roosevelt's initials.[96]

As outlined in the Villita Ordinance, the training center had two objectives: "First: fostering the production of arts and crafts which are useful, true in design, of good quality and workmanship, and which can be sold for a profit; and Second; the training of youth to fit into the modern complex life of a merciless machine age."[97] These goals situated La Villita within the vast tourist market of the American Southwest that popularized consumerism of Native arts and crafts. When Lou Block, NYA advisor for arts and crafts, lent his expert eye to the site in early 1940, he reported to NYA headquarters a positive evaluation of its training scheme: "I have been able to see here in actual practice a pattern for craft training for which I have always been very hopeful but had no way of proving."[98] The following year, he published an essay ruminating on the importance of training youth in creative arts. The piece heavily featured La Villita in praise of its expansive program that offered Mexican American youth, "whose occupational history had been more or less limited to pecan shelling or migrant labor in the Texas beet fields," greater economic opportunities in "largely uncompetitive avenues."[99]

An impressive list of other professional artists and craftspeople was called on to lend their expertise as either consultants or teachers at La Villita. To develop the arts and crafts program, the NYA hired Texan Blanding Sloan, a nationally known puppeteer, stage set designer, and instructor of crafts. Local artist Mary Vance Green, in charge of craft design, taught ceramics, weaving, blacksmithing, and sheet-metal workshops, while locally renowned ceramics artist Harding Black showed workers how to create bright red floor clay tiles and bake them in the kiln built on cite to mimic early Texas brickmaking methods. Figure 21 shows twenty-year-old Priscilla Camacho, second cousin of President Manuel Ávila Camacho of Mexico, modeling clay figures of borderland cowboys in La Villita's ceramics shop under Black's supervision.[100] Maverick also brought in his friend George Biddle, the renowned Philadelphia painter and muralist who had played a critical role in establishing the Federal Art Project, for a two-month artist-in-residence to observe and advertise various arts and crafts operations at both La Villita and the Witte Museum in San Antonio.[101] Meanwhile, the WPA's Arts and Crafts Division contributed to La Villita under the technical supervision of Ethel Wilson Harris, who owned three decorative art tile companies in San Antonio and operated a Mexican arts and crafts shop out of the restored granary at Mission San José.[102]

Maverick was keenly interested in the progress of the young NYA men and women employed at La Villita and visited the workshops frequently.[103] Because of his knowledge of how day-to-day work proceeded, the mayor often communicated directly with NYA staff with requests for more labor and materials for the project, sometimes taking complaints straight to state

FIGURE 21: Twenty-year-old Priscilla Camacho, second cousin of President Manuel Ávila Camacho of Mexico, models clay figures in La Villita's ceramics shop. Source: University of Texas at San Antonio Special Collections.

administrator Jesse Kellam or national administrator Aubrey Williams rather than the district director or onsite supervisor. He frequently grumbled about understaffing, insufficient quantities of materials, and long periods awaiting materials—complaints shared across the restoration projects, as discussed with the Dock Street Theatre and Henry Whitfield House. But Maverick also quibbled over minute decisions, such as the NYA's choice of lamps to install, which he called "monstrosities."[104] He also bemoaned the city's expenditures compared to the NYA's, which he felt had not followed through on what it originally agreed to fund. He reminded Kellam, "Technically, it may be merely a local job, but by recognition it is a national job." Kellam was sympathetic and assured Maverick that the NYA "guarantees completion of all building as planned at La Villita."[105]

Kellam might have called Maverick "the 'damndest' fellow I ever knew," but their professional and personal relationship grew stronger over the course of the restoration project.[106] Indeed, NYA district director Fenner Roth described Maverick as "kind of a sounding board for the [NYA] administration. I imagine he had about as much pull with them as anybody."[107] A good example of Maverick's rapport with the NYA is his communication with national administrator Aubrey Williams. Throughout the project, Maverick repeatedly nagged his old friend for not visiting the site, calling it "an *outrage*." In private communication to the "lazy bureaucrat" he even jokingly threatened to send a chaffing letter to President Roosevelt because of the NYA director's inattention to the Texan site.[108] Williams took the criticism in jest, as intended, and he wrote Maverick, "The town should be erecting a monument to you for getting things for them. Never has any half dozen lousy mayors gotten as much for them in six terms as you have in one."[109] The NYA director did eventually make his way to La Villita in August 1941 on a tour of southwestern states, stopping for lunch at the Cos House after visiting NYA projects at Alamo High School Stadium and Duncan Air Field Depo.[110]

Creating the Tourist Landscape of Romantic Spanish San Antonio

From the project's inception, Maverick cast La Villita's worth in both social and economic terms. At the same time that he described the "value of the project as a cultural center . . . impossible to estimate," he said it would be "worth $500,000 to San Antonio in salaries and increased real estate values."[111] Maverick told the city commission: "The development of San Antonio as a tourist center is of great importance. . . . *We have the best hotels and tourist quarters in America*. Let us fill them; let us offer clean amusements, parks, playgrounds, the zoo, sanitary and modern swimming pools, golf grounds, dude ranches, and our historic sites, which are the greatest in America."[112] Maverick jumped into an economic campaign already set in motion. In 1938, before he became mayor, the city administration initiated a project to develop the part of the San Antonio River that ran through the central business area. An improvement district composed of property owners along the river contributed $75,000 through bond issue before the WPA approved around $325,000, bringing the final cost of the project to over $400,000. As part of the so-called river beautification project, WPA workers floodproofed the river's Great Bend so a disastrous flood like the one of 1921, which killed over fifty people in the West Side and spurred the construction of the Olmos Dam, would not happen again.[113]

The river project developed the city's natural resource as a tourist site, business hub, and civic asset. Plans for the outdoor theater (shortly named the Arneson Theater), flower markets, food stalls, curio shops, and landscaped river banks earned the river project the nickname of "the Venice of America." Relief workers built public walks along which local artisans would sell their wares and installed floodlights to give a "moonlight effect."[114] The "cultural value" and "practical value" combined would make the River Walk, as it was later called, a boon to the developing tourist industry, clearly built around the idea of quixotic Spanish heritage.[115] The architect of the river design, Robert Hugman, noted that city advertising painted "the river as one element of the picturesque, romantic place 'where life is different,'" a point the local press emphasized repeatedly in its descriptions of the river's development.[116]

Despite their shared dream of an idealized Spanish San Antonio luring tourists downtown, Maverick and Hugman butted heads. Maverick strongly supported the river project since it, like La Villita, boosted San Antonio's tourism industry. However, he was overzealous in his efforts to control what happened. Hugman recalled that Maverick tried to get him to hire the mayor's relative as a landscape architect. More egregiously, Maverick diverted WPA funds earmarked for the river to his pet project, La Villita. When Hugman discovered this infraction, he brought it to the attention of the River Board and his attorney, but he was fired immediately and his involvement with the river project ended.[117]

Although Hugman lost Maverick's support, his plan for the San Antonio River was backed by the SACS, by then the most powerful local actor in heritage projects.[118] In 1924, Emily Edwards and Rena Maverick Green founded the SACS in an effort to save the 1858 Greek Revival Market House on Market Street from demolition as part of a street widening project. While they lost the Market House, their efforts set off a movement to preserve "all things characteristic of San Antonio, things of historic as well as aesthetic value—losing which, San Antonio loses local color and atmosphere." The goal became to save "Old Buildings, Documents, Pictures, Names, Natural Beauty, and anything admirably distinctive of San Antonio." By the end of the 1930s, the organization had saved the Alamo, four Spanish missions, and the Spanish Governor's Palace.[119] One of the SACS's greatest accomplishments was the reconstruction of the granary at San José Mission, which in 1941 became a state park and National Historic Site jointly administered by the Catholic Church, the Texas State Parks Board, and the National Park Service.[120]

Historian Charles Hosmer in his landmark work *Preservation Comes of Age* concluded that minus "a few harried husbands," the San Antonio preservation community "was a woman's world."[121] He echoed Ethel Wilson Harris, supervisor of WPA arts and crafts, who said that besides Maverick's efforts, historic preservation in 1930s San Antonio was "a women's movement" pushed forward "over the dead bodies of the men of San Antonio!"[122] Maverick astutely worked *with*, not against, the SACS, perhaps because his wife was a leading member but also because he recognized the organization's powerful authority in matters of cultural heritage. The president and one of the founding members of the organization, Amanda Cartwright Taylor, praised the mayor for his work at La Villita: "Already you have restored the uniquely gay bit of embroidery on the city's mantle—the chili stands. Now restoration of the old town helps to maintain the proper historical and geographical balance of the city."[123] Taylor's comment alluded to Maverick's decision shortly after assuming office to allow the "Chili Queens"—Mexican women who sold chili con carne and other popular Mexican dishes from street stands—to return to the streets of San Antonio with new sanitation regulations. This move earned the mayor favor with the SACS and helped curate a palatable Mexican American street identity.[124] Throughout her second term as SACS president from 1939 to 1941, Taylor worked with the city to promote La Villita, giving the mayor's project the SACS's meaningful stamp of approval and making use of the area's buildings for the organization's events.[125]

In addition to courting the women of the SACS, Maverick invited San Antonio businesspeople and other civic leaders to discuss the commercial possibilities of La Villita. Meeting in the restored Societies Building, the mayor and the river beautification committee jointly pitched the two projects as economic boons to the vacation business, which brought in $45 million a year, approximately twice as much as the cotton crop.[126] Both NYA state administrator Jesse Kellam and deputy administrator David Williams supported the plan, the latter of whom applauded La Villita for "bringing back the charm that made tourists come to San Antonio in the first place."[127] Maverick consequently launched a "publicity campaign" to make people aware of the "advantages of San Antonio as a tourist resort." He introduced a course for city employees "in the history of San Antonio and its desirability as a vacation spot" and had the city erect sixty-five arrow-shaped signs on arterial streets heading downtown to direct tourist traffic to La Villita.[128] Throughout the project, Maverick planned many open houses for residents and tourists alike to observe the work of NYA arts and crafts laborers. Over four hundred people attended one January afternoon open house in 1941, and Maverick himself conducted tours of the arts and crafts buildings every half hour, carrying a portable microphone so his audience could hear him describe the woodworking, metalcraft, and blacksmithing projects of the young artisans demonstrating their skills.[129]

Meanwhile, the Texas Writers' Project produced literature as part of the American Guide Series that beckoned visitors to enjoy the natural and historic features of the Alamo city, such as *Along the San Antonio River*.[130] The director of the city's Municipal Information Bureau also promoted La Villita by comparing it to the Rockefeller-funded restoration of Virginia's colonial capital. La Villita and its "humble homes of humble people" was to be a living historic village that nurtured active civic culture, rather than a restored city frozen in a moment of time, like Williamsburg.[131] In addition, the Maverick administration portrayed La Villita as an important site of civic education by incorporating the project into the public education system. The mayor sent information about the site to schools and invited classes to take tours. A marionette class from the San Antonio Vocational and Technical High School, home economics students from Texas University, and English classes from Brackenridge Senior School visited La Villita.[132] The president of the San Antonio Board of Education said that students, on observing the preservation of La Villita's historic buildings, "will better be able to imagine with what courage and ingenuity our forefathers solved the problems of their time."[133]

Further integrating preservation and tourism, the city promoted San Antonio as the gateway to Latin America with La Villita as its showpiece.

At the end of March 1940, the Mexican government's tourist commission and the Mexican Tourist Association organized a visit to San Antonio for more than forty travel editors of U.S. newspapers and representatives of Mexican tourism and auto associations. The event concluded with a supper at La Villita hosted by Maverick and city commissioners.[134] The mayor frequently used La Villita as a stage to welcome important Latin American figures, like the aide and brother of the president of Mexico, and twenty-high ranking military officials from nine Latin American countries. Upon receiving the "traditional key of generous hospitality," the Bolivian Army's chief of staff remarked, "With it we open our hearts and will keep it open for our brothers of the United States."[135] As a symbol of political goodwill, President Manuel Ávila Camacho of Mexico gifted an eighteen-foot "heroic-sized" statue of Father Hidalgo to the city, which was placed in La Villita's Juarez Plaza.[136]

Perhaps the most effective advertising of the restoration was its hosting of significant commemorative events and other festivities, thereby claiming a key geographic role in the city's public culture. Large crowds flocked to La Villita for events such as Guadalupe Day, the SACS's Yule pageant, Latin-American Day, New Citizens Day, and Civic Day, when more than eight thousand San Antonians inspected La Villita's workshops and Cos House while the WPA's Tipica Orchestra played on the patio.[137] In October 1940, the SACS's popular annual Indian Harvest Festival evolved into the River Jubilee in celebration of the San Antonio River beautification project. It included a water pageant at the Arneson Theater and a masquerade dance at La Villita.[138] During the following April's Fiesta Week, which celebrated the 1836 U.S. victory at the Battle of San Jacinto, La Villita hosted the royal ball for the first time. "King Antonio" and his royal company sailed a fleet of colorfully decorated boats carrying Mayor Maverick and city officials, army representatives, and Texas Cavaliers down to the river theater where they disembarked to attend the ball with spectators dressed in "Spanish, Mexican, cowboy, and similar costumes to add to the 'color'" of the event.[139]

While Anglo city officials and civic organizations largely created and promoted these events based on stereotyped ideas of Hispanic culture, some people of Mexican descent participated. For example, Mexican American singer Rosita Fernández, known as the "San Antonio Rose," began her sixty-year career giving live performances on the river during the 1930s tourism campaign. In inventing herself as Rosita, an "authentic representation of Mexico," Fernández became a "cultural bridge between Texas Mexicans and

FIGURE 22: Women pose in a sightseeing boat on the San Antonio River, with the Arneson River Theater in the background, 1942. Source: University of Texas at San Antonio Special Collections.

Texas Euro-Americans," argues historian Mary Ann Villareal.[140] It is also important to note that the mayor's administration did not focus its tourism efforts in the six-block area east of Santa Rosa Street that was the heart of the Mexican commercial quarter in San Antonio, with the highest concentration of businesses owned, operated, and patronized by Mexicans. Rather than use a "real" landscape to promote the Hispanic identity of the city, Maverick and other planners created idealized tourist landscapes in the form of La Villita and the River Walk that emphasized abstracted Spanish rather than authentic Mexican cultural roots (fig. 22).[141]

"Outstanding Shrine" or "Maverick's Folly"?

Both the Texas and central DC offices of the NYA indisputably considered La Villita a resounding success. NYA staff frequently visited La Villita, and other notable federal representatives toured the site on visits to San Antonio as well, including the director of the Tennessee Valley Authority and First Lady Eleanor Roosevelt.[142] As the project unfolded, the NYA's central office requested additional copies of the Villita Ordinance, the WPA booklet *Old Villita*, and the NYA progress report to send out to state offices. From the beginning, the NYA put considerable resources toward promoting the restoration, which it highlighted as a national model of the agency's outstanding historical work.[143]

For instance, in late 1939 the central office sent a representative from its Motion Picture Section to San Antonio to capture some color footage of La Villita while it was undergoing excavation and restoration. Cameraman Jimmie Lederer filmed shots of NYA youth engaged in construction work, local musician Eddie Martinez performing with his Latin-American band, "girls dancing and practicing their fandangoes," the "Chili Queens" serving meals, and Mayor Maury Maverick—the "daddy of La Villita"—signing and placing the official seal on the Villita Ordinance. The band recorded "El Rancho Grande" for the short film's soundtrack, "bringing out the point that as this project moves along, it will carry the romance of Old Mexico."[144] The clips culminated in a film called *NYA as a Good Neighbor*, clearly pushing the theme of pan-Americanism.[145]

The site received external positive attention as well; out-of-town visitors and presses in cities such as Chicago and Dallas sang the praises of the preservation activity occurring in the Texan city.[146] The *San Antonio Light* published a letter to the editor from a northern tourist spending the winter of 1941 in town. He and his wife visited the Alamo, the Spanish Governor's Palace, and Fort Sam Houston, but of all the city's historic sites, he wrote, La Villita "challenged my curiosity." The couple viewed the ceramics, woodcraft, and metal shops, and found over one hundred NYA youth workers to be "bright, happy, and intelligent . . . deeply interested in learning a craft that would fit them for a better life." La Villita, "now rising Phoenix-like into a new and permanent place in this city's list of outstanding shrines, is well worth the attention of all who would pause here for a while," he concluded.[147]

However, while NYA staff and visitors heaped praise on the project, some San Antonio residents responded with ambivalence. They clearly understood La Villita as the mayor's "pet project" whose "fame has been noised abroad."[148] Some townspeople and newspaper writers objected to the cost of the project, which they thought drained city resources and the coffers of the mayor's office in particular.[149] The fire chief recalled that "the newspapers were against Maury [Maverick] personally," and some locals referred to La Villita as "Maverick's Folly."[150] In the spring of 1940, the historically conservative *San Antonio Express* reported that the city's "side-stepping of relief problems" had become "more and more evident." The journalist argued that if city funds were available for La Villita and the river beautification project, they also should be available for hiring charity investigators, supporting the WPA sewing room project, and running the city council hospital.[151] This reproval echoed those leveled at the Dock Street Theatre by Charleston residents, who questioned whether New Deal funds should go toward the revival of a theater rather than the county jail.

The specific criticism of the mayor misspending funds also appeared in a "resume" of Maverick's first year in office published by *San Antonio Light* columnist Don Politico, which listed fifteen political missteps. Among the mayor's failure to reduce taxes, frequent travels that took him away from city hall, reorganization of the police department, and the annexation of suburbs, three of Maverick's faults were directly related to La Villita. They included the "promiscuous use of advertising tax funds," "a propensity for employing many out-of-town experts," and "constant concentration in public statements and expenditures upon La Villita."[152]

Thus, critics of the site explicitly linked the NYA project with Maverick's broader local reform agenda, and public disapproval escalated in the lead up to the mayoral election of 1941. Former mayor C. K. Quin, head of the San Antonio Democratic machine and longtime Maverick rival, made overspending city funds and preoccupation with La Villita central to his anti-Maverick campaign.[153] Maverick defended his fixation on La Villita because it was a major source of revenue to the city and "built up good will in Latin America."[154] Indeed, the sitting mayor repeatedly used La Villita as a symbol of his political activity on behalf of Mexican Americans, pointing to its creation of jobs for a thousand youth, "a majority of them Mexicans." Five hundred, he said, later found employment with private firms because of their NYA training.[155]

Nonetheless, the Quin ticket exploited Maverick's weakness as a racial reformer and took up the old refrain of his harboring communist sympathies. This strategy played on both anticommunist and strong anti-Mexican sentiments in San Antonio because of the link between Mexican union workers and the local Communist Party.[156] Despite Maverick's liberal reforms, San Antonio during the late Depression years remained segregated—relief projects were separated by race, for example—as well as antiunion, which did not help improve the economic situation of Mexican Americans or Blacks.[157] Partly because the Quin campaign made promises of more patronage jobs for Black voters, Maverick was unable to defeat the Democratic machine for a second time, and he lost in a runoff election with 19,799 votes to Quin's 20,982.[158] Despite his overtures of aiding Mexican Americans, Maverick lost much of the West Side vote he had received in the 1939 election.[159] Maverick never sat in public office again but soon took on important roles during the war years on the War Production Board and as chair of the Smaller War Plants Corporation.

With Maverick's defeat in 1941, the city council voted to transfer La Villita to the parks department, which then placed the project under the control of the recreation department. Just three days after the election, Maverick held a dedication for the incomplete La Villita, his "favorite child." Ceremonies in front of the new, still-unfinished Bolivar Building included speeches by the mayor-elect Quin, NYA state administrator Jesse Kellam, and the city park commissioner, who now oversaw the project.[160] A few weeks into his term, Quin returned to inspect La Villita. He reportedly found the work "highly commendable," and he committed the city to finishing the restoration, despite attacking the project during his campaign (fig. 23).[161]

While carrying out Maverick's project, the new mayor found a way to diminish his predecessor's presence at the site, moving the statue of Father Hidalgo gifted to Maverick by the president of Mexico from Juarez Plaza to a less prominent position over a mile away.[162] It is doubtful, however, that anyone forgot who championed the restoration. Maverick was easily elected the first president of the La Villita Corporation, a nonprofit organization chartered in April 1940. He held that position until 1946 and simultaneously served on the La Villita Advisory Board.[163] Once formed, the La Villita Corporation under Maverick's leadership was incredibly proactive, keeping scheduled NYA work on task and weighing in on matters like the construction of a bell tower and the addition of toilets to the San Martin House.[164]

Later that summer, to celebrate the almost-finished restoration, local NYA officials sponsored a formal dinner at the Cos House. At the dinner, NYA national administrator Aubrey Williams said, "We are proud of the La Villita, nationally proud," and praised the federal agency for "making work redignified" by teaching youth to learn skills with their hands and bodies.[165] In early February 1942, a little behind schedule, all construction work came to an end, and in a final ceremony at City Hall, Mayor Quin formally received

FIGURE 23A: Bird's-eye view of La Villita from the Smith-Young Tower before the restoration began in July 1939. Source: University of Texas at San Antonio Special Collections.

FIGURE 23B: Bird's-eye view of La Villita from the Smith-Young Tower in May 1941 when the Bolivar Building was still under construction. Source: University of Texas at San Antonio Special Collections.

the project along with a small-scale model of the restored village made by NYA workers.[166]

A New Wartime Role for La Villita

Although La Villita as a construction project had been brought to a close, the NYA training center at the site continued to operate and took on heightened significance as American involvement in World War II increased. Even before the United States joined the war in December 1941, workers at La Villita began producing items for national defense programs. Since the project's beginning, Maverick had explained that "half of the training at La Villita should be along industrial lines, work which can be used for governmental or military purposes." While still mayor, he lent the city's radio equipment to the NYA and expressed his wish that La Villita would become a community center for soldiers training in San Antonio.[167] In March 1941, the Board of Trustees offered the use of La Villita to the Federal Security Agency's military recreational department and began hosting a monthly dance series throughout 1941 cosponsored by the city and the San Antonio Recreation Council, with assistance from the WPA's soldiers service bureau.[168]

As the war ramped up, NYA production transitioned to making supplies for army camps, and in late December, Mayor Quin offered the use of all La Villita buildings to the Bexar County Chapter of the American Red Cross, which offered Motor Corps driving lessons and nursing classes for women throughout the war. The Bolivar Building's intended use as a library and museum was postponed; when it opened in early 1942, it was as a training center for first-aid and other wartime workers.[169] La Villita became an important and visible site of Latin American friendship during the war and earned the praise of high-ranking Washington figures, including the undersecretary of state, the chief of staff of the War Department, and the secretary of the Department of Agriculture.[170] Under the auspices of the Office of the Coordinator of Inter-American Affairs, a wartime agency created to promote pan-Americanism, the National Press Club led tours of the site for newspaper representatives from Central and South America in 1943 in support of hemispheric cooperation.[171]

During the war years, the board voted to increase rental fees and service charges for use of the Cos House, Juarez Plaza, Bolivar Building, and the Arneson Theater, which kept La Villita solvent.[172] After the war ended, the board rededicated itself to ensuring La Villita was "used for the purposes for which it was intended—the development of the Arts and Crafts, especially as related to Texas and Pan-American culture." As a steward of public property, it could not let La Villita become "a public burden," which meant operating La Villita "on a business basis" to bring in sufficient income. It is clear there was local interest in continuing local arts and crafts businesses on site; for example, the La Villita Corporation accepted a proposal to establish "La Villita Weavers," a school of instruction for handweaving, in the old NYA weaving shop rent free for one year, after which the board would then receive a "fair percentage of profit or rental basis."[173] Soon after, the board gave a ceramics center a trial of six months. In addition to achieving economic profitability, the board sought to expand the geographical boundaries of La Villita. The original ordinance creating the site in 1939 mandated that "the physical scope of the area shall be increased as time goes on," and, accordingly, over the next twenty years the city acquired land one block to the east of King Philip Street, which included such properties as St. Philips College, the Dashiell House, and the McAllister House and Mercantile.[174]

By the mid-1940s, La Villita had earned recognition not only for its strategic political and economic position as the "gateway" to Latin America but as a charming historical and tourist destination, attracting attention from companies and publications across the nation. In 1943, the Magnolia Petroleum Company included the San Antonio site in its annual calendar, the March issue of the *Woman's Home Companion* featured spring fashion models posing with La Villita in the background, and the *Ladies' Home Journal* ran a feature article with photographs.[175] For San Antonio itself, La Villita remained a popular place for festivities; one resident recalled, "There was something going at La Villita every night, in 2 or 3 places."[176] The Alamo Chapter of the Daughters of the American Revolution and the SACS sponsored free concerts performed by school bands in the summers, and several local clubs held regular meetings at the Cos House throughout the year, including the San Antonio Historical Society and the Pan-Hellenic Society.[177]

In the early 1950s, former mayor Maury Maverick, still serving on the La Villita Advisory Board, reflected on the most significant project of his mayoral administration, which he said had succeeded beyond his wildest dreams. He praised the "self-sustaining pay-as-you-go feature," which he called "amazing for a public governmental project," acknowledging the economic success of his New Deal federal-local initiative. In the early years of the Cold

War, employing the political rhetoric of the period, he summarized La Villita's influence as a skills development center, educational historic site, and symbol of freedom:

> We are proud of Villita's achievements, and the hundreds of thousands of people it has served. Villita has taught young people trades, and how to work. It is a truly American institution, and a symbol of the best in American life and government. We are particularly pleased that practically every school or college student of San Antonio visits Villita once or more a year, while additional tens of thousands of students and pupils visit from all over America. . . . Villita is a great teacher and symbol of free government, against the totalitarian forces in the world today.[178]

In similar vein, the chair of the advisory board called La Villita "a training center for American citizenship and the American way of life, not a mere Recreation Center."[179] What began as a project to position San Antonio's Mexican American youth for economic success through skills development became a training ground for American democracy. The irony of this shift is that the group the project initially targeted to aid were increasingly othered as "illegal aliens" and discriminated against in housing policies and employment practices as La Villita assumed this citizenship-building role.

The "Final Report" of the NYA program in Texas highlighted La Villita as "one of the most outstanding arts and crafts projects in Texas and possibly the nation."[180] Maverick had thought so too. While working on a manuscript about La Villita in the spring of 1941, he asked his friend Texas historian Walter Prescott Nobb to provide edits. Nobb suggested Maverick remove "moral observations" in reference to the New Deal. Maverick very humbly accepted other criticisms of his writing but disagreed with Nobb's objection to "injecting philosophy" into a book about La Villita, for, he wrote, "Villita is a New Deal agency itself."[181] He understood that his pet project symbolized the kind of public infrastructure local governments were capable of building in their cities through New Deal cooperative arrangements. Despite their disagreements during the project, architect La Villita architect O'Neil Ford called Maury Maverick "the only person in Texas at that time that would put up money and put up a fight for preservation and conservation of historic values."[182] While women members of civic organizations like the San Antonio

Conservation Society had been enshrining historic sites in San Antonio for decades, Ford's comment speaks to Maverick's critical role as a political figure in activating city and federal resources for cultural heritage projects, articulating a powerful and positive goal for preservation to improve San Antonians' quality of life.

Conclusion

The Cavalcade of American history is recorded in the nation's historic shrines. Each identifies some memorable event—adds a page to the inspiring Saga of our Nation's advancement. . . . All drew attention from these cradles of democracy that preserve America's history. Daughters of [the] American Revolution, a few private organizations, these endeavor to preserve our hallowed shrines. The Federal Government extends a helping hand. Works Progress Administration Projects daily restore our heritage to its Colonial glory. Decayed structures are restored to their former beauty. Dilapidated buildings are repaired. Old public meetinghouses are renovated. Battle sites that gave our liberty are reproduced. Thus a richer tradition is insured for posterity.

"Cavalcade," n.d., Works Progress Administration

In December 1937, Ellen S. Woodward, director of the Division of Women's and Professional Projects of the WPA, published "Alles in Aims: The Story of What America Is Doing to Preserve Its Historical Heritage" in the magazine of the Daughters of the American Revolution. Regarding the general "neglect of historical landmarks and historical records" preceding the Depression years, Woodward commended the federal government for aiding in the "preservation of valuable historical treasures" as part of its work relief program. With the support of "allies" like the Daughters of the American Revolution, new federal initiatives rallied members of local and state governments, historical societies, preservation organizations, and other patriotic groups to prevent the loss of more historic material as "modern modes replace the old ways of living."[1] Supported financially and ideologically by the New Deal, Americans took action to preserve historic landmarks as emblems of the nation's progress. The restored shrines stood as reminders of the United States' great trials and triumphs, and served as symbols to inspire future generations to invent, experiment, and envision better tomorrows.

In December 1941, four years after Woodward published her article, the United States entered World War II and New Deal projects, including the historic shrine restorations, gradually came to an end. Wartime production and military roles assuaged the unemployment that had plagued the nation for the past decade, and the government terminated relief programs over the course of the war years. The Works Progress Administration, which had been renamed the Work Projects Administration after a government reorganization in 1939, ended on June 30, 1943. The National Youth Administration, established within the WPA, continued until January 1, 1944, with training centers like La Villita reoriented toward the war effort.

Altogether, the New Deal programs put in place by the Roosevelt administration undeniably alleviated some of the economic and environmental hardships of the Great Depression through an unprecedented bureaucracy of federal work relief. The Civilian Conservation Corps paved thousands of miles of roadways, planted trees, and expanded national and state park systems; the Tennessee Valley Authority built hydroelectric dams that brought electricity to vast areas of the South; the Agricultural Adjustment Administration assisted farmers dealing with inflation and drought; and the Federal Housing Administration helped people secure affordable homes through mortgage lending. The WPA, which alone provided employment for 8.5 million Americans during its eight years, stimulated the economy by creating construction jobs to build civic structures, parks, airports, roadways, schools, and more. The NYA, meanwhile, addressed problems afflicting the nation's youth and trained over 4.7 million young men and women in employable skills through public works and service projects.

But just as the economic and political infrastructure of the United States required radical visions for stabilization after the stock market crash of 1929, so too did the cultural sector of the country demand creative and innovative solutions. The WPA fostered an unparalleled level of federal support for the arts through the four programs of Federal One and promoted historical and cultural endeavors in its public works projects. The agency's historic shrine restorations, categorized as construction projects and pushed forward by local government officials, boosters, and civic organizations, helped rebuild America's landmarks from the bricks, stone, and wood that bore witness to some of the nation's most important historical events. By highlighting the restorations in official publications, the WPA's Division of Information took advantage of the popularity of these locally conceived restorations and put them into service to aid the agency's broader message of recovery through employment in vital nation-building work.

Although the New Deal programs came to a close during World War II, the four historic shrines discussed in this work all found success in the decades following their restorations and continue to serve their communities today. In Charleston, South Carolina, the Dock Street Theatre, the most expensive of the four projects, reclaimed the southern city's role as a regional center of theatrical activity when its doors reopened in November 1937. By creatively engineering the reappearance of the eighteenth-century theater, Charlestonians with a stake in maintaining political and cultural control reinvented the architectural character of the southern city's historic core. The federal project also advanced the political career of Charleston mayor Burnet Maybank, whose skillful use of New Deal funds and strong relationship with leading figures in the Roosevelt coalition helped him win the governorship in 1938. The Dock Street Theatre remained a segregated space throughout the Jim Crow period and experienced some financial troubles in the mid-twentieth century. However, the site has undergone subsequent restorations and remains a central cultural institution in a city whose rich history, still prominently illustrated in its antebellum buildings, continues to drive its thriving tourism industry. The theater is now home to the resident professional theatre company Charleston Stage and is a popular host venue of many of the city's leading performing arts and cultural festivals, including the annual Spoleto Festival and Charleston Literary Festival.

In the small New England town of Guilford, Connecticut, the Henry Whitfield State Museum once again welcomed visitors to learn about the heritage bequeathed by its Puritan forebearers when the WPA restoration came to a close in 1937. By restoring the three-hundred-year-old Old Stone

House as accurately as possible to its original 1640 appearance, the museum's Board of Trustees and architect J. Frederick Kelly hoped that the sturdy stone foundations of the town's earliest English settlers would serve as a reminder of perseverance through difficult times. Within the broader context of the colonial revival movement, these local leaders positioned the museum as a bulwark against modern changes wrought by increased industrialization, urbanization, and immigration. Facing criticism from the town's traditional memory keepers, the project also demonstrated resistance to the expanded role of professionals and outsiders in historical commemoration. When Democratic governor Wilbur Cross narrowly lost reelection in 1938 and the state returned to Republican control, the restored shrine stood as an expression of the state's commitment to conservative Yankeedom. The Henry Whitfield House remains the state museum today, and, as of the time of writing, staff are in the process of revising its historical interpretation to more fully contextualize the impact of settler colonialism on Connecticut's history.

Moving to central Minnesota, the home and land of the state's one-time favorite native son, Charles A. Lindbergh, became a premier historical and nature destination for residents and tourists alike to enjoy in the 1930s. The Lindbergh State Park WPA project centered on the famous flier but also endorsed New Deal conservation policies and public recreation in natural areas, and represented remarkable cooperation between political parties and government offices. The state park's association with one of America's leading celebrities, its focus on environmental issues in a Farmer-Labor Party–controlled state, and its smaller scale in terms of necessary staffing and supplies kept the site in good standing in the WPA offices (local, state, and national), the state legislature, and the town of Little Falls. Shortly after the park project was completed, however, the Farmer-Labor Party lost control of the state, and public opinion of Charles Lindbergh became divided as he assumed a leading role in the isolationist America First Committee. Yet, these political changes did not impact the popularity of the park throughout the mid- to late twentieth century, and it has remained a frequented vacation site where visitors can reaffirm their relationship with the natural world. Since expanded, the over five-hundred-acre Lindbergh State Park, managed by the Minnesota Department of Natural Resources, has endured as an attractive recreational facility with many of the WPA-built structures, including the park shelter and water tower, still standing. The Charles Lindbergh House and Museum, composed of the aviator's boyhood home and a visitor center, is administered by the Minnesota Historical Society and interprets Lindbergh's family, fame, and complicated legacy.

Finally, in San Antonio, Texas, Mayor Maury Maverick understood historic preservation as a political instrument, using the restoration of La Villita to improve sanitation and housing conditions, train disadvantaged youth in employable skills, revive local arts and crafts production, boost tourism, and foster goodwill among Pan-American allies as the United States moved toward involvement in a global war. At the same time, the restoration displaced residents, mostly Mexicans and Mexican Americans, and created a romanticized version of that very group's history and culture. Yet, La Villita offers more than just a story of elite Anglo historical commemoration. Architect O'Neil Ford's vision ensured that La Villita endorsed a more inclusive vision of whose past mattered in its emphasis on vernacular nineteenth-century structures. Moreover, Mexican American youth acquired construction and arts skills in La Villita's training programs, which they applied to government defense industry work during the war and private employment afterward. Throughout the rest of the twentieth century and into the next, La Villita continued to expand its physical boundaries as the city of San Antonio purchased surrounding historic properties, twenty-seven of which are listed in the National Register of Historic Places' La Villita Historic District. Although not a site of craft production today, La Villita's retail shops sell local artisans' wares, and the area remains a popular destination for tourists visiting the nearby River Walk, developed simultaneously with La Villita as a WPA project.

The successful federal-local cooperative preservation of the New Deal, although temporary, laid the groundwork for expanded federal investment in historic preservation in the post–World War II period. While the federal preservation program was fairly dormant in the two decades following the end of the war, it grew exponentially in the 1960s. Scholars and practitioners widely recognize the National Historic Preservation Act of 1966 as transformative legislation that strengthened national preservation policy and more formally regulated the government's stewardship of the nation's cultural resources. The act created the National Register of Historic Places, introduced the review process for federal agencies known as Section 106, authorized a grant fund that became the Historic Preservation Fund, mandated the appointment of State Historic Preservation Officers, and established the Advisory Council on Historic Preservation, all measures that continue to guide professional practice today. Subsequent decades saw the growth of national and state historic tax credit programs, founded in 1976, and the continued expansion of the National Park Service, which maintains the Secretary of the Interior's Standards for the Treatment of Historic Properties within its over four hundred units. The blueprints for this increased federal activity were created during the oft-overlooked New Deal period in histories of historic preservation.

This book conceives of historic preservation as an expression of civic duty as it sits at the intersection of historical education, entertainment, tourism, business, and politics. Although centered on historical narratives embodied in old places, preservation is an outgrowth of plans for the future and consequently, as we have seen in these shrine restorations, a manifestation of political ambitions. That the success story of diverse preservation examples operating under an overarching federal umbrella took place during the Great Depression is not insignificant. Amid the largest economic crisis in U.S. history, some Americans channeled their energy, their dreams, their professional talents, their leisure time, and their physical and intellectual labor into maintaining built legacies of the American past. With money tight, the political landscape unstable, and another global war looming on the horizon, people eagerly applied for federal dollars to be siphoned into historic property maintenance. As part of the New Deal apparatus, historic preservation offered jobs for unskilled and skilled laborers, encouraged the professionalization of the field, and developed local tourism. To that end, New Deal preservation speaks to the benefits of investment in a cultural economy by the state.

Although the American preservation movement today is strongly supported by federal, state, and local programs and financial incentives, it is important that we revisit chapters of its history to better understand *how* and *why* Americans across time and space managed their contemporary political milieus through manipulations of historic places, exploring both the pragmatic means and ideological motivations that made cooperative endeavors successful. The 1930s–1940s shrine restorations offer valuable insights into the mindsets of Americans dealing with the catastrophic fallout of economic disaster and the political strategies they employed to cope in a period of immeasurable change. During the Depression era, politicians, architects, craftspeople, and relief workers in diverse places and with different constraints and resources used historic preservation to shape both historical memory and contemporary cultural and political identities. We should see their success as indicative of what can happen when grassroots meet grasstops advocacy in campaigns to protect our built heritage. The combination of community leaders, influential politicians, and supportive federal programs resulted in the safeguarding of local landmarks that have continued to be centerpieces of community and historical memory in each of the four locations examined.

While this book highlights examples of productive federal-local partnerships to fund cultural endeavors, it also illuminates how privilege and political power shaped those efforts, to both negative and positive outcomes. The exclusion of community groups with stakes in the projects (such as Black Americans in Charleston and Mexican Americans in San Antonio) placed limitations on the extent to which the shrine restorations were truly democratic in political nature and inclusive with regard to historical narrative. Furthermore, in ignoring darker or more difficult chapters in American history, such as slavery, Native genocide, and the exploitative labor practices of industrial capitalism, New Deal actors missed opportunities to engage with the legacy of challenging historical events and ideologies and the histories of oppressed communities, which could have offered alternative and productive comparisons for their own struggles.

Preserving historic places has been, and continues to be, a means to reckon with the "American Way of Life," but our twenty-first-century approach can and should apply a more discerning lens with regard to inclusion of race, gender, and class differences than during the Depression years. Custodianship of historic places matters as a form of cultural production and as a system of exhibiting our collective values and political priorities. It presents a potentially democratic way for people to tangibly claim space, to ensure the continuation of particular narratives and insert new voices and new stories into the public historical landscape.

Thus, the New Deal preservation projects explored herein offer significant lessons for the preservation field. For one, they communicate the importance of inclusion in political processes of preservation; diverse voices and perspectives must be welcomed on city councils, architecture and historical commissions, and institutional boards. In her pivotal study on the power of place in urban environments, Dolores Hayden explained that "a politically conscious approach to urban preservation must go beyond the techniques of traditional architectural preservation . . . to reach broader audiences. It must emphasize public processes and public memory."[2] Public agencies involved in preservation, including historic review or landmark commissions, housing authorities, and art councils, are beholden to their constituents, who as taxpayers and the audience of civic projects have equal investment in their outcomes.

The New Deal preservation efforts also inform us of the need to advocate for increased access to federal dollars for municipal governments to fund preservation initiatives. While federal and local actors clearly used preservation as an economic driver of historical tourism in the 1930s–1940s, we are reimagining the possible roles historic preservation can play in addressing contemporary issues facing our communities today, such as the green building and sustainability movement, projects of decolonization, and affordable housing. As we have seen during the Depression, historic and civic infrastructure improved residents' quality of life, and enhanced federal support can strengthen preservation's power as an ameliorative political and social force in the present.

Ultimately, the New Deal restorations exemplify the essential understanding that historic preservation is political and can mobilize ideas and people. My hope is that this book serves as a reminder to readers that we all are historical actors who can employ historic preservation to make the changes we want to see in our communities. Understanding the context of past efforts helps inform us of our political and social responsibilities as current practitioners and encourages us to interrogate our own circumstances in which we make political and spatial choices about protecting and interpreting historic places. Writing about broader commemorative impulses, public historian Seth C. Bruggeman reminds us that "the way we remember today—no matter how original it may seem—perpetuates ideas and values from the past that may or may not nourish our hopes for the future. . . . A memory put in motion tends to stay in motion. Commemoration has the power to accelerate certain memories while diverting others."[3] Studies of historical commemoration, like this one, call for self-reflexivity on the part of all players involved in historical memory-making to question *which* historical narratives we promote and *why*. We must ask ourselves whether the places we protect and the memories they hold still resonate with contemporary values.

It is clear that historic places take on psychological importance and that place attachment and historical memory inform our civic and political lives.[4] During the Depression, Americans well understood that local landmarks provide both a sense of rootedness and inspiration for building better futures. Places feel more permanent than ephemeral cultural forms in their tangibility, physicality, and immovability, and in a particularly turbulent period Americans sought to stake a claim somewhere. They chose historic buildings and what they considered inspiring narratives of the past that helped them understand who they were and, importantly, who they could become. This framing of the preservation impulse is essential because it calls attention to the proactive and forward-looking capacity of preservation work.

However, American society should not need to be wracked by a severe economic depression and shock to its cultural identity to utilize preservation as a regenerative political and cultural force. And, indeed, we see many communities across the United States employing preservation methods to stimulate economic revitalization, insert marginalized voices into narratives from which they have been excluded, and adopt sustainable architecture practices. Much as federal actors and local constituents chose in the 1930s and 1940s when managing the multitudinous crises of the Great Depression, we too can choose preservation as a tool for improving our national and local lives. In imagining what that system might look like, we can engage in important discussions of ethics (especially when it comes to funding sources), sustainability, balancing the historical record, and redressing historical economic inequities.

Appendix A

"Historic Shrines" Restored by the Works Progress Administration

These projects are drawn from the WPA's Division of Information, A1 764, Records Concerning the Restoration of Historic Shrines, 1937–38, National Archives at College Park (although some projects included in the files postdate 1938). The projects are listed by state (with city in parentheses) and alphabetized by project name. This is not a complete list of all the historic sites restored with help from the WPA but the most commonly referenced in the agency's "shrines" material. Additionally, I have not included general archaeological surveys; the sites in agency literature identified only as "Indian Mounds"; natural features, like the Constitutional Elm (Corydon, Indiana); and most of the monuments and memorials restored by the WPA, of which there were at least fifty. The monuments I have listed received frequent mention in communication among WPA staff and administrators. The intention of including this appendix is to demonstrate the geographical and historical diversity of the projects, as well as their limitations; for example, notice the absence of any sites primarily associated with Black history, whereas Native American history is well represented through both archaeological and architectural restoration work.

Most of these projects are included in the Living New Deal digital project administered by the Department of Geography, University of California, Berkeley, https://livingnewdeal.org/. The website also documents projects that involved historic preservation activity funded by other agencies including the National Youth Administration, the Civilian Conservation Corps, and the Public Works Administration.

Arizona
Fort Lowell (Tucson)
Globe Ruins (Globe)
Old Fort Misery and Sharlot Hall Museum (Prescott)
Pueblo Grande Museum (Phoenix)
Tuzigoot Pueblo (near Clarkdale)

California
Adobe Chapel (San Diego)
Asistencia Mission San Gabriel (San Bernardino)
Fort Humboldt (Eureka)
General Mariano Guadalupe Vallejo Estate (Sonoma)
Presidio of San Francisco, Officers' Club (San Francisco)
Telegraph Hill (San Francisco)
Vallecito Stage Station (Agua Caliente Springs)

Colorado
Fort Bent [Bent's Old Fort] Museum (La Janta)
Fort Vasquez (Platteville)
Simpson's Rest (near Trinidad)

Connecticut
Fairfield Town Hall (Fairfield)
Henry Whitfield State Museum/Old Stone House (Guilford)

Delaware
Old Arsenal (New Castle)
Old Court House (New Castle)

Florida
City Gates (St. Augustine)
Fort Jefferson (Key West)

Georgia
Cyclorama "Battle of Atlanta" (Atlanta)
Fort Hawkins (Macon)
Indian Village (St. Simons Island)
Liberty Hall/Alexander Stephens's Home (Crawfordville)
Ocmulgee National Monument (Macon)

Illinois
Fort Dearborn (Chicago)
Graue Mill (DuPage County)
Old Market House (Galena)

Indiana
Lincoln Pioneer Village (Rockport)
Mounds State Park (Anderson)
Tippecanoe Battlefield (Lafayette)

Iowa
Fort Atkinson (Winneshiek County)

Kentucky
Dr. Ephraim McDowell House (Danville)

Louisiana
American Legion Home (New Orleans)
The Cabildo (New Orleans)
Fort Pike (New Orleans)
Jackson Barracks (New Orleans)
Judah P. Benjamin House (Belle Chasse)
Mansfield Battlefield (De Soto Parish)
Napoleon House (New Orleans)
Old Ladies Home (New Orleans)
Old State Capitol of Louisiana (Baton Rouge)
Parish Court House (Clinton)
Upper Pontalba Building (New Orleans)

Maine
Fort Knox (Prospect)

Maryland
Allegany County Academy (Cumberland)
Carroll Mansion (Baltimore)
Flag House (Baltimore)
Peale Museum (Baltimore)

Massachusetts
Abbott Hall (Marblehead)
Arbella [ship] (Salem)
Boston Common (Boston)
Charlestown Navy Yard (Boston)
Christ Church Burying Ground (Cambridge)
Faneuil Hall (Boston)
Fort Banks (Winthrop)
Fort Independence (Boston)
Fort Revere (Hull)
Fort Sewall (Marblehead)
Fort Warren (Boston Harbor)
Hartshorne House (Wakefield)
Old Dog Town Common (Gloucester)
Old Grist Mill (Eastham)
Old Power House Park (Somerville)
Old State House (Boston)
Old Town Hall (Salem)
Old Town House (Marblehead)
Pulpit Rock (West Bridgewater)
Quincy Market (Boston)
Springfield Arsenal Barracks (Springfield)
Watertown Arsenal (Watertown)

Michigan
Fort Holmes (Mackinac Island)
Fort Wayne (Detroit)

Minnesota
Faribault House (Faribault)
Hudson Bay Company Stockade (Grand Portage)
Lindbergh Home and State Park (Little Falls)
Sibley House (Mendota)

Missouri
Arrow Rock State Park (Saline County)
Fort D (Cape Girardeau)

New Hampshire
Historic Quarter at Portsmouth (Portsmouth)

New Jersey
Abbott Farm (near Trenton)
Dey Mansion (Passaic)
Grover Cleveland Birthplace (Caldwell)
Indian King Tavern (Haddonfield)
Morristown National Historic Park (Morristown)
Rockingham House (Kingston)
Somers Mansion (Somers Points)
Trent House (Trenton)
Von Steuben House (New Bridge)
Walt Whitman Home (Camden)

New Mexico
Chaco Canyon National Memorial (San Juan County)
Coronado State Monument (Sandoval County)
Fort Marcy (Santa Fe)
Paako Ruins (San Pedro)
Palace of Governors (Santa Fe)

New York
Buffalo Museum of Natural Sciences (Buffalo)
Conference House (Staten Island)
Fort Niagara (near Youngstown)

Fort Schuyler (New York City)
Gracie Mansion (New York City)
Grant's Tomb (New York City)
Jumel Mansion (New York City)
Rochester Municipal Museum (Rochester)
Stark's Knob (near Saratoga)
Statue of Liberty (New York City)
Stony Point Battlefield Museum (Stony Point)
Sir William Johnson Hall and Fulton County Jail (Johnstown)

North Carolina
Federal Mint Museum (Charlotte)
Fort Raleigh (Roanoke Island)
John Wright Stanly House (New Bern)
Nathanial Macon House (Warrenton)

North Dakota
Fort Abercrombie (Abercrombie)
Fort McKeen/Abraham Lincoln (Mandan)
On-a-Slant Indian Village (near Mandan)

Ohio
Benjamin R. Hanby House (Westerville)
Borglum's Northwest Territory Monument (Marietta)
Fort Amanda (Lima)
Fort Recovery State Memorial (Mercer County)
Fort St. Clair (Eaton)
Ulysses S. Grant Birthplace (Point Pleasant)
William Henry Harrison Memorial (North Bend)
Wood County Historical Museum (Bowling Green)

Oklahoma
Fort Gibson (Fort Gibson)
Sequoyah's Home/Cabin (Sallisaw)

Oregon
Amaton Springs (Champeog)

Pennsylvania
Fort Le Beouf (Washington)
Fort Washington Park (Montgomery County)
Greene County Court House (Waynesburg)
Independence Hall (Philadelphia)
League Island Park (Navy Yard) (Philadelphia)
Reval Cemetery (Bedford)
Revolutionary War Memorial Park (Bedford)
Robert E. Peary Memorial (Cresson)
Schuylkill Arsenal (Philadelphia)
Singley House (Philadelphia)
Thaddeus Stevens Blacksmith Shop (Fayetteville)
Trinity High School (Washington)
Village Hall (Montrose)
Vernon-Wister House (Philadelphia)

Puerto Rico
Ballajá Barracks/Cuartel de Ballajá (San Juan)
Casa Blanca (San Juan)
Castillo San Cristóbal (San Juan)
Castillo San Felipe del Morro (San Juan)

Rhode Island
Benefit Street Armory (Providence)
Fort Independence (Robin Hill)
Narragansett Indian Village at Roger Williams Park (East Greenwich)

South Carolina
Dock Street Theatre (Charleston)
Winthrop College Chapel (Rock Hill)

South Dakota
Dinosaur Park (Rapid City)
Fort Sisseton (Lake City)

Tennessee
Fort Loudon (Vonore)

Fort Negley (Nashville)
The Hermitage (Nashville)

Texas
Camp Colorado (Coleman)
Fort Belknap (Newcastle)
Mission San José (San Antonio)
San Jacinto Battleground (Harris County)

Utah
Dinosaur National Monument (Uintah County)

Vermont
Georgia Town Hall (Georgia)

Virginia
Christopher Johnson Cottage (Lynchburg)
Valentine Museum (Richmond)

Washington
Fort Nisqually (Tacoma)

Wisconsin
Aztalan Indian Village (Jefferson County)
Nelson Dewey Homestead at Wyalusing State Park [formerly Nelson Dewey
    State Park] (Bagley)
Villa Louis/Dousman Estate (Prairie du Chien)

Wyoming
Fort Bridger (Minta County)
Fort Caspar (Natrona County)
Medicine Bow Petrified Forest and Dinosaur Bed (Carbon County)

Appendix B

National Park Service
Restoration Policies, May 19, 1937

General Restoration Policy:

The motives governing these activities are several, often conflicting: aesthetic, archeological and scientific, and educational. Each has its values and its disadvantages.

Educational motives often suggest complete reconstitution, as in their heyday, of vanished, ruinous, or remodeled buildings and remains. This has often been regarded as requiring removal of subsequent additions, and has involved incidental destruction of much archeological and historical evidence, as well as of aesthetic values arising from age and picturesqueness.

The demands of scholarship for the preservation of every vestige of architectural and archeological evidence—desirable in itself—might, if rigidly satisfied, leave the monument in conditions which give the public little idea of its major historical aspect or importance.

In aesthetic regards, the claims of unity or original form or intention, of variety of style in successive periods of building and remodeling, and of present beauty of texture and weathering may not always be wholly compatible.

In attempting to reconcile these claims and motives, the ultimate guide must be the tact and judgment of the men in charge. Certain observations may, however, be of assistance to them:

(1) No final decision should be taken as to a course of action before reasonable efforts to exhaust the archeological and documentary evidence as to the form and successive transformations of the monument.

(2) Complete record of such evidence, by drawings, notes and transcripts should be kept, and in no case should evidence offered by the monument itself be destroyed or covered up before it has been fully recorded.

(3) It is well to bear in mind the saying: "Better preserve than repair, better repair than restore, better restore than construct."

(4) It is ordinarily better to retain genuine old work of several periods, rather than arbitrarily to "restore" the whole, by new work, to its aspect at a single period.

(5) This applies even to work of periods later than those now admired, provided their work represents a genuine creative effort.

(6) In no case should our own artistic preferences or prejudices lead us to modify, on aesthetic grounds, work of a bygone period representing other artistic tastes. Truth is not only stranger than fiction, but more varied and more interesting, as well as more honest.

(7) Where missing features are to be replaced without sufficient evidence as to their own original form, due regard should be paid to the factors of period and region in other surviving examples of the same time and locality.

(8) Every reasonable additional care and expense are justified to approximate in new work the materials, methods and quality of old construction, but new work should not be artificially "antiqued" by theatrical means.

(9) Work on the preservation and restoration of old buildings requires a slower pace than would be expected in new construction.

Battlefield Area Restoration Policy:

Consideration of a proper restoration policy for historical areas raises many important problems. Not the least of these is the proper application of such a policy to national battlefield areas. Those areas offer conditions not usually present in other historical sites and the problem is more immediate in view of the present rapid development program.

In a sense a wise policy might better be described as one of stabilization rather than restoration. Stabilization embraces necessary restoration without subordinating to it the entire physical development program.

It is convenient to discuss the problem in two parts, the elements usually presented in a battlefield area when the National Park Service takes it over, but before any development program has been initiated; and, the successive steps in a sound stabilization program.

I. When the National Park Service takes over a military area, it usually consists of the following elements:

A. What was there when the battle was fought, including evidences of the battle, such as earthenworks, cleared fields, ruined foundations, etc.

B. Subsequent additions, including forest growth, modern buildings, monuments, and markers. Some of these subsequent additions, such as the intrusions of unsightly and modern structures, have been injurious to the appearance of the area. Other additions, however, have improved it. For example, forest growth of 75 years frequently is a desirable witness to the age and the dignity of a battlefield area and fortifies the impression upon those visiting the area.

II. To stabilize conditions on a battlefield area after it is taken over, the following policies are hereby approved:

A. Undesirable modern encroachments on the battlefield scene shall be eliminated as soon as practicable. Not everything that has occurred since the battle can be considered an encroachment. Obviously, modern structures and intrusions which have been due to other than natural conditions and which introduce a jarring note rather than contribute to the normal accretions of age are the elements which should be eliminated. These include modern buildings, high-speed highways, gas stations, transmission lines, and other obviously incongruous elements. Normal forest growth, the natural changes of stream channel, the operation of other natural processes which seem destined never to be controlled, should not be eliminated.

B. Having eliminated undesirable encroachments, those features of the area which hamper a clear understanding of the engagement also should be eliminated. For example, where forest growth has obstructed an important vista or where a road location conveys a mistaken notion of troop movements, that feature should be modified or eliminated for educational reasons.

C. Restoration, which seems advisable to aid understanding and to restore the natural landscape for clearing and naturally representing the battlefield area, should be made as funds therefore are obtained. Such restorations may be made for structures, earthworks, plant growth, etc. It is recognized that, in each case of restoration, there is present a danger of introducing an artificial element into what had been previously a natural scene. Natural processes should be allowed to operate and dignify with age the natural scene.

The foregoing policies should aid in developing a battlefield area to provide a combination of elements remaining from the time of the battle, plus the normal additions of age affected through the natural accretion of natural processes. When a battlefield area has been so treated as to represent this combination, it can be said to be "stabilized."

Sample Restoration Policy:

The Advisory Board approves the guiding policy of the treatment of the Morristown camp site, in accordance with which the restoration of only a very small number of representative structures is attempted, and expresses its opposition to any attempt at complete or large-scale restoration of such sites, especially where the building of structures is involved.

Source: Harlan D. Unrau and George F. Williss, *Administrative History: Expansion of the National Park Service in the 1930s* (Denver: Denver Service Center, National Park Service, 1983), appendix 2, "Restoration Policies, May 19, 1937, Development of Restoration and Preservation Policies: 1935–1941," https://www.nps.gov/parkhistory/online_books/unrau-williss/adhia2.htm.

Notes

Introduction

1. The letter asked state administrators to report projects previously undertaken by other emergency work programs that predated the WPA, including the Civil Works Administration and the Federal Emergency Relief Administration. Some restorations began as projects of one of these earlier programs before transitioning to the WPA. Other specific examples named in the letter included the Ephraim McDowell House in Danville, Kentucky; the Benjamin Hanby House in Westerville, Ohio; Fort Niagara in Youngstown, New York; and the Dinosaur Park in Rapid City, South Dakota. The state administrators received a follow-up letter in October 1937 requesting supplementary information. David K. Niles to State Works Progress Administrators, "Information Service Letter No. 34," February 15, 1937; "Information Service Letter No. 42," October 22, 1937, Box 1, Folder "Historic Shrines—Overview," Records Concerning the Restoration of Historic Shrines, 1937–1938 (hereafter Entry A1 764), Records of the Work Projects Administration, Record Group 69 (hereafter RG 69), National Archives at College Park, College Park, MD (hereafter NACP).

2. Norman Fitts to Carl Wennerblad, September 18, 1937, Box 6, Folder "HS—Massachusetts—General Correspondence," Entry A1 764, RG 69, NACP.

3. Other possible titles for the booklet included "America's Cavalcade," "America Marches On!," "America Through the Centuries," "Remembrance of Things Past," and "Landmarks of Liberty." Norman Fitts to Paul Ellerbe, Internal Memorandum, May 18, 1937, Box 1, Folder "290A—Correspondence; 'Conference re Booklet on Historic Shrines,'" n.d., Box 1, Folder "HS 39 (Info on Illustrated Book on Restoration Projects)," Entry A1 764, RG 69, NACP.

4. Works Progress Administration, *Inventory: An Appraisal of the Results of the Works*

Progress Administration (Washington, DC: U.S. Government Printing Office, 1938), 33–34.

5. "America's Cavalcade," n.d., Box 1, Folder "Historic Shrines," Entry A1 764, RG 69, NACP. For discussion on the role of posterity in motivating historical commemoration, see Nick Yablon, *Remembrance of Things Present: The Invention of the Time Capsule* (Chicago: University of Chicago Press, 2019).

6. "Historic Military Shrines Restored by U.S.," September 21, 1939, Box 15, Folder "290-B Historical Shrine Reconstruction and Preservation," Information Service (Primary File), 1936–42 (hereafter Entry A1 678), RG 69, NACP.

7. The New Deal work relief programs often promoted the trope of the "American Way," positioning political democracy and economic security as the pillars of American society. William W. Bremer, "Along the 'American Way': The New Deal's Work Relief Programs for the Unemployed," *Journal of American History* 62 (December 1975): 636–52; Wendy L. Wall, *Inventing the "American Way": The Politics of Consensus from the New Deal to the Civil Rights Movement* (New York: Oxford University Press, 2008).

8. "Historic Shrine Summary," April 18, 1939, Box 15, Folder "290-B Historical Shrine Reconstruction and Preservation," Entry A1 678, RG 69, NACP.

9. Robert D. Leighninger Jr., *Long-Range Public Investment: The Forgotten Legacy of the New Deal* (Columbia: University of South Carolina Press, 2007), 9, 200.

10. James T. Patterson, *The New Deal and the States: Federalism in Transition* (Westport, CT: Greenwood Press, 1981), 166.

11. Thomas Kessner, "Fiorello H. LaGuardia and the Challenge of Democratic Planning," in *The Landscape of Modernity: New York City, 1900–1940*, ed. David Ward and Olivier Zunz (Baltimore: Johns Hopkins University Press, 1992), 321. See also Mark I. Gelfand, *A Nation of Cities: The Federal Government and Urban America, 1933–1965* (New York: Oxford University Press, 1975).

12. Leighninger, *Long-Range Public Investment*, 200.

13. "Cavalcade," n.d. Box 1, Folder "Historic Shrines—Overview," Entry A1 764, RG 69, NACP. The WPA also funded the Monument Restoration Project of the New York City Parks Department between 1934 and 1937, which began as a Public Works of Art Project and was overseen by sculptor Karl Gruppe. I did not count this project's monuments in the nearly two hundred shrine projects. Monuments restored under this project include the Washington at Valley Forge Memorial in Brooklyn, the William Cullen Bryant Memorial in Bryant Park, the General Sherman Sculpture and Soldiers and Sailors Memorial Arch in Manhattan's Grand Army Plaza, among many others. The monuments are documented in the Living New Deal digital project administered by the Department of Geography, University of California, Berkeley, https://livingnewdeal.org/.

14. "WPA Aid Sought for Grant's Tomb Repairs: Group Urges Move for 1939 World's Fair," *New York Times*, May 30, 1937.

15. "Indiana . . . Spencer County, Three Best Projects (Sheet 2)," September 1936, Box 1, Folder "290A—Correspondence," Entry A1 764, RG 69, NACP. For more on

the reconstruction of Lincoln Pioneer Village in the 1930s, see Keith A. Erekson, *Everybody's History: Indiana's Lincoln Inquiry and the Quest to Reclaim a President's Past* (Amherst: University of Massachusetts Press, 2012).

16. "Old House to Wake and Live," *Philadelphia Inquirer*, July 21, 1937.

17. My approach of examining each site as both a physical place and an idea is informed by Heidi Aronson Kolk, *Taking Possession: The Politics of Memory in a St. Louis Town House* (Amherst: University of Massachusetts Press, 2019), 16.

18. Neil M. Maher, *Nature's New Deal: The Civilian Conservation Corps and the Roots of the American Environmental Movement* (New York: Oxford University Press, 2008), 7.

19. John Bodnar, *Remaking America: Public Memory, Commemoration, and Patriotism in the Twentieth Century* (Princeton, NJ: Princeton University Press, 1994), 14.

20. For explanations of the four preservation techniques employed by contemporary practitioners in the field—preservation, restoration, rehabilitation/adaptive use, and reconstruction—see Anne E. Grimmer, *The Secretary of the Interior's Standards for the Treatment of Historic Properties with Guidelines for Preserving, Rehabilitation, Restoring and Reconstructing Historic Buildings* (Washington, DC: U.S. Technical Preservation Services, National Park Service, Department of the Interior, 2017).

21. Works Progress Administration, *Inventory*, 33–34.

22. Jennifer McLerran examines how romantic primitivism prevalently shaped the relationship between New Deal art program administrators and Native artists in *A New Deal for Native Art: Indian Arts and Federal Policy, 1933–1943* (Tucson: University of Arizona Press, 2009).

23. For some works that explore the ways marginalized communities, and Black Americans in particular, activated New Deal programs to their political, economic, and cultural advantages, see Lauren Rebecca Sklaroff, *Black Culture and the New Deal: The Quest for Civil Rights in the Roosevelt Era* (Chapel Hill: University of North Carolina Press, 2009); Patricia Sullivan, *Days of Hope: Race and Democracy in the New Deal Era* (Chapel Hill: University of North Carolina Press, 1996); and Kate Dossett, *Radical Black Theatre in the New Deal* (Chapel Hill: University of North Carolina Press, 2020).

24. William F. McDonald, *Federal Relief Administration and the Arts: The Origins and Administrative History of the Works Project Administration* (Columbus: Ohio State University Press, 1969); Sharon Ann Musher, *Democratic Art: The New Deal's Influence on American Culture* (Chicago: University of Chicago Press, 2015); Barbara Dianne Savage, *Broadcasting Freedom: Radio, War, and the Politics of Race, 1938–1948* (Chapel Hill: University of North Carolina Press, 1999); Sklaroff, *Black Culture and the New Deal*; Dossett, *Radical Black Theatre in the New Deal*; Nicholas Natanson, *The Black Image in the New Deal: The Politics of FSA Photography* (Knoxville: University of Tennessee Press, 1992); Cara A. Finnegan, *Picturing Poverty: Print Culture and FSA Photographs* (Washington, DC: Smithsonian Institution Press, 2003); Barbara Melosh, *Engendering Culture: Manhood and Womanhood in New Deal Public Art and Theater* (Washington, DC: Smithsonian Institution Press, 1991).

25. Although the Tennessee Valley Authority's dam projects are most recognizable for their architectural modernism, TVA architects integrated local geography, history, and traditions into their designs. See chapters by Jan Wolff, "Redefining Landscape," and Todd Smith, "Almost Fully Modern: The TVA's Visual Art Campaign," in *The Tennessee Valley Authority: Design and Persuasion*, ed. Tim Culvahouse (New York: Princeton Architectural Press, 2007); Karal Ann Marling, *Wall-to-Wall America: A Cultural History of Post-Office Murals in the Great Depression* (Minneapolis: University of Minnesota Press, 1982); Marlene Park and Gerald E. Markowitz, *Democratic Vistas: Post Offices and Public Art in the New Deal* (Philadelphia: Temple University Press, 1984); and Benjamin Filene, *Romancing the Folk: Public Memory and American Roots Music* (Chapel Hill: University of North Carolina Press, 2000).

26. Leighninger, *Long-Range Public Investment*.

27. Jason Scott Smith, *Building New Deal Liberalism: The Political Economy of Public Works, 1933–1956* (New York: Cambridge University Press, 2006), 166.

28. Phoebe Cutler, *The Public Landscape of the New Deal* (New Haven, CT: Yale University Press, 1985), 8.

29. Marguerite S. Shaffer, *See America First: Tourism and National Identity, 1880–1940* (Washington, DC: Smithsonian Books, 2001); Shaffer, "Seeing the Nature of America: The National Parks as National Assets, 1914–1929," in *Being Elsewhere: Tourism, Consumer Culture, and Identity in Modern Europe and North America*, ed. Shelley Baranowski and Ellen Furlough (Ann Arbor: University of Michigan Press, 2001), 155–84; Michael Berkowitz, "A 'New Deal' for Leisure: Making Mass Tourism during the Great Depression," in Baranowski and Furlough, eds., *Being Elsewhere*, 185–212. See also Mordecai Lee, *See America: The Politics and Administration of Federal Tourism Promotion, 1937–1973* (Albany: State University of New York Press, 2020).

30. David J. Nelson, *How the New Deal Built Florida Tourism: The Civilian Conservation Corps and State Parks* (Gainesville: University Press of Florida, 2019).

31. Michael Holleran, *Boston's "Changeful Times": Origins of Preservation and Planning in America* (Baltimore: Johns Hopkins University Press, 1998); Charles B. Hosmer Jr., *Preservation Comes of Age: From Williamsburg to the National Trust, 1926–1949*, 2 vols. (Charlottesville: University Press of Virginia, 1981); James M. Lindgren, *Preserving Historic New England: Preservation, Progressivism, and the Remaking of Memory* (New York: Oxford University Press, 1995); Randall Mason, *The Once and Future New York: Historic Preservation and the Modern City* (Minneapolis: University of Minnesota Press, 2009); Max Page, *The Creative Destruction of Manhattan, 1900–1940* (Chicago: University of Chicago Press, 1999).

32. Whitney Martinko, *Historic Real Estate: Market Morality and the Politics of Preservation in the Early United States* (Philadelphia: University of Pennsylvania Press, 2020). See also Martinko, "Progress and Preservation: Representing History in Boston's Landscape of Urban Reform, 1820–1860," *New England Quarterly* 83, no. 2 (June 2009): 304–34.

33. James M. Lindgren, for example, wrote that the Society for the Preservation of New

England Antiquities' modern management of preservation in the first two decades of the twentieth century "ultimately led to the founding of such organizations as the NTHP [National Trust for Historic Preservation] and state preservation offices," ignoring the expanded and bureaucratic state in New Deal preservation efforts. Lindgren, *Preserving Historic New England*, 11. Diane Barthel attributed the activism of professionals in the government and private sector leading to the founding of the NTHP to the "localism that had characterized American preservation from the beginning," which "made coordination very difficult." Barthel, *Historic Preservation: Collective Memory and Historical Identity* (New Brunswick, NJ: Rutgers University Press, 1996), 22. Other important works of preservation scholarship that either ignore or devote minimal critical engagement with the nuances of 1930s New Deal preservation include Robert Stipe, ed., *A Richer Heritage: Historic Preservation and the Twenty-First Century* (Chapel Hill: University of North Carolina Press, 2003); William Murtagh, *Keeping Time: The History and Theory of Preservation in America*, 8th ed. (1998; Hoboken, NJ: John Wiley, 2010); and Max Page and Randall Mason, eds., *Giving Preservation a History: Historic Preservation in the United States*, 2nd ed. (2003; New York: Routledge, 2019).

34. Lynne M. Calamia, "A New Deal for Historic Preservation: The Impact of Relief on the Cultural Landscape of Preservation" (PhD diss., Pennsylvania State University, 2015).

35. Cathy Stanton, *The Lowell Experiment: Public History in a Postindustrial City* (Amherst: University of Massachusetts Press, 2006); Seth C. Bruggeman, *Lost on the Freedom Trail: The National Park Service and Urban Renewal in Postwar Boston* (Amherst: University of Massachusetts Press, 2022).

36. Dolores Hayden, *The Power of Place: Urban Landscapes as Public History* (Cambridge, MA: MIT Press, 1995), 12.

37. Patricia West, *Domesticating History: The Political Origins of America's House Museums* (Washington, DC: Smithsonian Institution Press, 1999); Barthel, *Historic Preservation*; Karal Ann Marling, *George Washington Slept Here: Colonial Revivals and American Culture, 1876–1986* (Cambridge, MA: Harvard University Press, 1988); Kirk Savage, *Monument Wars: Washington, D.C., the National Mall, and the Transformation of the Memorial Landscape* (Berkeley: University of California Press, 2009); Denise D. Meringolo, *Museums, Monuments, and National Parks: Toward a New Genealogy of Public History* (Amherst: University of Massachusetts Press, 2012); Seth C. Bruggeman, ed., *Born in the U.S.A.: Birth, Commemoration, and American Public Memory* (Amherst: University of Massachusetts Press, 2012); Erekson, *Everybody's History*.

38. I am indebted to the analytical concept of a "sense of place" in framing my understanding of historic preservation as political memory work. David Glassberg, *Sense of History: The Place of the Past in American Life* (Amherst: University of Massachusetts Press, 2001). For works that explore the politics of placemaking, see Ned Kaufman, *Place, Race, and Story: Essays on the Past and Future of Historic Preservation* (New York: Routledge, 2009), and Daniel Bluestone, *Buildings, Landscapes, and Memory: Case Studies in Historic Preservation* (New York: Norton, 2011).

39. Seth C. Bruggeman, "Introduction," in *Commemoration: The American Association for State and Local History Guide*, ed. Seth C. Bruggeman (Lanham, MD: Rowman and Littlefield, 2017), 4.

40. Some works that examine the politics of historical memory-making, especially through the lens of place, include Hayden, *The Power of Place*; Stephanie E. Yuhl, *A Golden Haze of Memory: The Making of Historic Charleston* (Chapel Hill: University of North Carolina Press, 2005); Lewis F. Fisher, *Saving San Antonio: The Precarious Preservation of a Heritage* (Lubbock: Texas Tech University Press, 1996); Susan T. Falck, *Remembering Dixie: The Battle to Control Historical Memory in Natchez, Mississippi, 1865–1941* (Jackson: University Press of Mississippi, 2019); W. Fitzhugh Brundage, ed., *Where These Memories Grow: History, Memory, and Southern Identity* (Chapel Hill: University of North Carolina Press, 2000); Paul A. Shackel, ed., *Myth, Memory, and the Making of the American Landscape* (Gainesville: University Press of Florida, 2001); Seth C. Bruggeman, *Here, George Washington Was Born: Memory, Material Culture, and the Public History of a National Monument* (Athens: University of Georgia Press, 2008); Bluestone, *Buildings, Landscapes, and Memory*; and Bruggeman, *Lost on the Freedom Trail*.

41. Bluestone, *Buildings, Landscapes, and Memory*, 17.

42. Warren Susman, *Culture as History: The Transformation of American Society in the Twentieth Century* (New York: Pantheon Books, 1984), 197; Susman, ed., *Culture and Commitment, 1929–1945* (New York: George Braziller, 1973); Michael Denning, *The Cultural Front: The Laboring of American Culture in the Twentieth Century* (New York: Verso, 1997). For more works on "radical" or leftist culture and the Popular Front, see Bill Mullen and Sherry Lee Linkon, eds., *Radical Revisions: Rereading 1930s Culture* (Urbana: University of Illinois Press, 1996); Richard H. Pells, *Radical Visions and American Dreams: Culture and Social Thought in the Depression Years* (New York: Harper & Row, 1973); Lizabeth Cohen, *Making a New Deal: Industrial Workers in Chicago, 1919–1939* (Cambridge: Cambridge University Press, 1990); Robert Cohen, *When the Old Left Was Young: Student Radicals and America's First Mass Student Movement, 1929–1941* (New York: Oxford University Press, 1993); and Ronald D. Cohen, *Depression Folk: Grassroots Music and Left-Wing Politics in 1930s America* (Chapel Hill: University of North Carolina Press, 2016).

43. Lawrence W. Levine, *The Unpredictable Past: Exploration in American Cultural History* (New York: Oxford University Press, 1993), 222.

44. Victoria Grieve, *The Federal Art Project and the Creation of Middlebrow Culture* (Urbana: University of Illinois Press, 2009), 6.

45. Bruggeman, "Introduction," 4.

Chapter 1

1. John Dos Passos, *The Ground We Stand On: Some Examples from the History of a Political Creed* (New York: Harcourt, Brace, Jovanovich, 1941), 3.

2. The term "usable pasts" was coined by literary critic Van Wyck Brooks in a 1918 essay in which he suggested that American history can be a source of cultural creativity

rather than an inhibition. Brooks, "On Creating a Usable Past," *The Dial*, April 11, 1918, 337–41.

3. Susman, "Introduction," in Susman, ed., *Culture and Commitment*, 2. See also Susman, *Culture as History*.

4. For the documentary impulse of the 1930s, see William Stott, *Documentary Expression and Thirties America* (Oxford: Oxford University Press, 1976).

5. Calamia, "A New Deal for Historic Preservation," 158. For an overview of the early historic preservation movement, see Charles B. Hosmer Jr., *Presence of the Past: A History of the Preservation Movement in the United States before Williamsburg* (New York: G. P. Putnam's Sons, 1965).

6. Lindgren, *Preserving Historic New England*.

7. Mason, *The Once and Future New York*; Page, *The Creative Destruction of Manhattan*.

8. For example, in 1913 the General Assembly and governor of Pennsylvania created the Pennsylvania Historical Commission under the jurisdiction of the Department of Public Instruction, with "the duty of marking and preserving the antiquities and historic landmarks of Pennsylvania." The commission mostly aided local groups' efforts to preserve and commemorate historic sites. Calamia, "A New Deal for Historic Preservation," 50.

9. Franklin D. Roosevelt, "Annual Message to Congress," January 4, 1935, The American Presidency Project, https://www.presidency.ucsb.edu.

10. For discussion of the religious tradition of tourism in the nineteenth century, see John F. Sears, *Sacred Places: American Tourist Attractions in the Nineteenth Century* (Amherst: University of Massachusetts Press, 1999). Pilgrimages to birthplaces of "great men" in particular helped develop the notion of historical site as shrine. See Bruggeman, ed., *Born in the U.S.A.*

11. West, *Domesticating History*, 1–37. For more on women's preservation efforts in the nineteenth century, see Gail Lee Dubrow and Jennifer B. Goodman, eds., *Restoring Women's History through Historic Preservation* (Baltimore: Johns Hopkins University Press, 2003), particularly part 1, "Documenting the History of Women in Preservation."

12. Sears, *Sacred Places*, 7, 9.

13. Devin C. Manzullo-Thomas, "Sacred Subjects: Religion and Commemoration in America," in Bruggeman, ed., *Commemoration*, 92. For more on sacred sites, particularly battlefield grounds, see Edward T. Linenthal, *Sacred Ground: Americans and Their Battlefields*, 2nd ed. (1991; Urbana: University of Illinois Press, 1993).

14. Hilary Iris Lowe, "Authenticity and Interpretation at Mark Twain's Birthplace Cabins," in Bruggeman, ed., *Born in the U.S.A.*, 99; Paul Reber and Laura Lawfer Orr, "Stratford Hall," in Bruggeman, ed., *Born in the U.S.A.*, 119.

15. Michael Kammen, *Mystic Chords of Memory: The Transformation of Tradition in American Culture* (New York: Knopf, 1991), 175, 295.

16. Anders Greenspan, *Creating Colonial Williamsburg: The Restoration of Virginia's Eighteenth-Century Capital*, 2nd ed. (2002; Chapel Hill: The University of North Carolina Press, 2009), 10.

17. For works that explore the popularity and architecture of the colonial revival movement, see Alan Axelrod, ed., *The Colonial Revival in America* (New York: Norton, 1985); David Gebhard, "The American Colonial Revival in the 1930s," *Winterthur Portfolio* 22, nos. 2–3 (Summer–Autumn 1987): 109–48; Richard Guy Wilson, Shaun Eyring, and Kenny Marotta, eds., *Re-Creating the American Past: Essays on the Colonial Revival* (Charlottesville: University of Virginia Press, 2006); and Marling, *George Washington Slept Here*.

18. The Colonial Dames of America established the Van Cortlandt Mansion, the Daughters of the American Revolution created the Betsy Ross House Museum, the Concord Woman's Club founded the Orchard House, and Carolina Emmerton started the Seven Gables Settlement Association to restore the House of Seven Gables. West, *Domesticating History*, 39–91.

19. West, *Domesticating History*, 45, 95–96.

20. Lindgren, *Preserving Historic New England*, 4, 99.

21. Yuhl, *A Golden Haze of Memory*; Fisher, *Saving San Antonio*; Falck, *Remembering Dixie*.

22. Bluestone, *Buildings, Landscapes, and Memory*, 248.

23. Shaffer, "Seeing the Nature of America," 157.

24. McLerran, *A New Deal for Native Art*, 18n32.

25. William Butler, "Another City upon a Hill: Litchfield, Connecticut, and the Colonial Revival," in Axelrod, ed., *The Colonial Revival in America*, 15.

26. For discussion of the federal government's promotion of national parks in the interwar period, see Shaffer, "Seeing the Nature of America," 155–84, 157 (quotation).

27. Shaffer, *See America First*, 4, 142.

28. Wendy Griswold, *American Guides: The Federal Writers' Project and the Casting of American Culture* (Chicago: University of Chicago Press, 2016), 80.

29. Daniel Bluestone explores the development of Virginia's historical highway markers in the 1920s and 1930s in the chapter "Drive-by History: Virginia's Historic Highway Marker Program," in *Buildings Landscapes, and Memory*, 240–55, 241 (quotation).

30. Cutler, *The Public Landscape of the New Deal*, 57.

31. Laurence Vail Coleman later served as the director of the American Association of Museums beginning in 1943. Coleman, "Historic Houses Hang Out Latchstrings," *New York Times*, March 11, 1934.

32. Charles B. Hosmer covers both sites in his essential *Preservation Comes of Age*. For comprehensive works on Colonial Williamsburg, see Richard Handler and Eric Gable, *The New History in an Old Museum: Creating the Past at Colonial Williamsburg* (Durham, NC: Duke University Press, 1997), and Greenspan, *Creating Colonial Williamsburg*. Greenfield Village's history is explored in depth in Jessie Swigger, *"History Is Bunk": Assembling the Past at Henry Ford's Greenfield Village* (Amherst: University of Massachusetts Press, 2014). Other industrialists were active in recreating historical America, most notably Albert O. Wells, head of American Optical Company, whose antique collecting led to the creation of Old

Sturbridge Village in Sturbridge, Massachusetts, in 1946, and Henry Francis du Pont of Delaware's industrialist du Pont family, who in the 1920s began amassing a collection of historic American decorative arts in his family home, eventually opening the Winterthur Museum in 1951.

33. Ford first opened Greenfield Village on October 21, 1929, just eight days before the stock market crash launched the Great Depression, although tours were given only on request at the time. Swigger, *"History Is Bunk,"* 1.

34. Ford also restored the farmhouse in Dearborn where he grew up to its appearance in 1876 when he was thirteen years old. David Glassberg, "History and the Public: Legacies of the Progressive Era," *Journal of American History* 73, no. 4 (March 1987): 974.

35. Swigger, *"History Is Bunk,"* 32.

36. William B. Rhoads, "The Colonial Revival," 2 vols. (PhD diss., Princeton University, November 1974), 1:524–25.

37. Swigger, *"History Is Bunk,"* 3–4.

38. Marling, *George Washington Slept Here*, 286.

39. For the "animated textbook" quote, see Norman Tyler, Ted J. Ligibel, and Ilene R. Tyler, *Historic Preservation: An Introduction to Its History, Principles, and Practice* (New York: Norton, 2009), 35. For further discussion of Ford's creation of Greenfield Village, see Kammen, *Mystic Chords of Memory*, 351–58; Marling, *George Washington Slept Here*, 283–90; Hosmer, *Preservation Comes of Age*, 1:75–97; and Barthel, *Historic Preservation*, 38–39.

40. Steven Conn, *Museums and American Intellectual Life, 1876–1926* (Chicago: University of Chicago Press, 1998), 154.

41. As Karal Ann Marling put it, "When Henry Ford moved buildings willy-nilly to Michigan, he effectively wiped away the context and meaning once determined by their sites and associations, and rendered them featureless." Marling, *George Washington Slept Here*, 290.

42. Many outdoor history museums opened across the United States in the 1940s and 1950s, including the Farmers' Museum in Cooperstown, New York (1944); Old Sturbridge Village in Massachusetts (1947); Old Salem in North Carolina (1950); Historic Deerfield in Massachusetts (1952); and Plimoth Plantation in Massachusetts (1957). Jessie Swigger, "Outdoor History Museums," The Inclusive Historian's Handbook, June 11, 2019, https://inclusivehistorian.com.

43. Greenspan, *Creating Colonial Williamsburg*, 16.

44. Kammen, *Mystic Chords of Memory*, 200.

45. Greenspan, *Creating Colonial Williamsburg*, 8, 12; Swigger, *"History Is Bunk,"* 26.

46. Greenspan, *Creating Colonial Williamsburg*, 41, 54.

47. Kenneth Chorley, 1941, quoted in Kammen, *Mystic Chords of Memory*, 374n89. For examination of "historical truth" as presented at Colonial Williamsburg and how interpretation has changed over time, see Handler and Gable, *The New History in an Old Museum*.

48. Jihong Kim and Bong Hee Jeon, "Restoration of a Historic Town to Commemorate

National Identity: Colonial Williamsburg in the Early Twentieth Century," *Journal of Asian Architecture and Building Engineering* 11, no. 2 (2012): 247.

49. W. A. R. Goodwin to John D. Rockefeller Jr., July 8, 1937, quoted in Greenspan, *Creating Colonial Williamsburg*, 55.

50. Greenspan, *Creating Colonial Williamsburg*, 32–34, 48.

51. As Anders Greenspan notes, "Since most Americans' perception of colonial history in the 1930s closely mirrored that presented by the restoration and did not encompass a broader understanding of the lives of laborers, African Americans, and women, there was little criticism directed at the restoration's presentation of the past." Greenspan, *Creating Colonial Williamsburg*, 43.

52. Swigger, "Outdoor History Museums."

53. This "cultural democracy" comprised, according to historian Jane De Hart Mathews, three elements: "cultural accessibility for the public, social and economic integration for the artist, and the promise of a new national art." Mathews, "Arts and the People: The New Deal Quest for a Cultural Democracy," *Journal of American History* 62, no. 2 (September 1975): 325.

54. Musher, *Democratic Art*, 27–28; Roy Rosenzweig and Barbara Melosh, "Government and the Arts: Voices from the New Deal Era," *Journal of American History* 77, no. 2 (September 1990): 596–608; Alfred Haworth Jones, "The Search for a Usable American Past in the New Deal Era," *American Quarterly* 23, no. 5 (December 1971): 710–11, 719–22.

55. The Section of Fine Arts work was completed (and preceded) by the Public Works of Art Project (PWAP), the first large-scale federal art project established in the Treasury Department with funds from the Civil Works Administration in December 1933. During its five months in operation, the PWAP employed 3,749 artists, who created 15,663 works of art to decorate public buildings and parks. A similar program, the Treasury Relief Art Project, employed artists to decorate 2,500 federal buildings between 1935 and 1939. Rosenzweig and Melosh, "Government and the Arts," 597.

56. For discussion of regionalism in New Deal art programs, see Lauren Kroiz, *Cultivating Citizens: The Regional Work of Art in the New Deal Era* (Oakland: University of California Press, 2018), and Jonathan Harris, *Federal Art and National Culture: The Politics of Identity in New Deal America* (Cambridge: Cambridge University Press, 1995). For studies of regionalism and the New Deal more broadly, see Robert L. Dorman, *Revolt of the Provinces: The Regionalist Movement in America, 1920–1945* (Chapel Hill: University of North Carolina Press, 1993), and Michael Steiner, "Regionalism in the Great Depression," *Geographical Review* 73 (October 1983): 430–46.

57. According to Sharon Ann Musher, more than one-third of the murals (415 of 1,116) had historical content and more than sixty dealt exclusively with history. Musher, *Democratic Art*, 95–96. For more on the Treasury Section of Fine Art murals, see Virginia M. Mecklenburg, *The Public as Patron: A History of the Treasury Department Mural Program* (College Park: Department of Art, University of Maryland, 1979);

Karal Ann Marling, *Wall-to-Wall America*; and Park and Markowitz, *Democratic Vistas*.

58. To produce the American Guide Series the FWP worked in conjunction with other federal agencies such as the NPS, CCC, U.S. Travel Bureau, and the National Travel Advisory Board. McLerran, *A New Deal for Native Art*, 20; Griswold, *American Guides*, 111.

59. Christine Bold, *The WPA Guides: Mapping America* (Jackson: University Press of Mississippi, 1999), 27.

60. Jerrold Hirsch, *Portrait of America: A Cultural History of the Federal Writers' Project* (Chapel Hill: University of North Carolina Press, 2003), 71.

61. Roderick Seidenberg to Joseph Gaer, September 28, 1938; Seidenberg to James G. Dunton, July 7, 1937, cited in Hirsch, *Portrait of America*, 73.

62. Henry G. Alsberg, "The American Guide," n.d., cited in Hirsch, *Portrait of America*, 48.

63. Hirsch, *Portrait of America*, 91–92.

64. Bold, *The WPA Guides*, xvi.

65. Hirsch, *Portrait of America*, 102–3.

66. Lewis Mumford, "Writers' Project," *New Republic* 92 (October 20, 1937): 306–7.

67. Griswold, *American Guides*, 31–32. See also Bold, *The WPA Guides*, 4–6, and Andrew S. Gross, "The American Guide Series: Patriotism as Brand-Name Identification," *Arizona Quarterly: A Journal of American Literature, Culture, and Theory* 62 (2006): 86. Gross specifically looks at the guidebooks for California and Arizona. For discussion of how the WPA countered boondoggling criticism between 1935 and 1938, see Smith, *Building New Deal Liberalism*, 135–59.

68. Griswold, *American Guides*, 10.

69. Berkowitz, "A 'New Deal' for Leisure," 203.

70. Lee, *See America*, xii, 142.

71. Berkowitz, "A 'New Deal' for Leisure," 203, 194. This article provides an excellent account of how the expansion of paid vacations in the decade following World War I and joint federal and local "promotional apparatus for tourism" turned tourism into a mass phenomenon.

72. Harlan D. Unrau and G. Frank Williss, "To Preserve the Nation's Past: The Growth of Historic Preservation in the National Park Service during the 1930s," *Public Historian* 9, no. 2 (Spring 1987): 25, 28.

73. Harlan D. Unrau and G. Frank Williss, *Administrative History: Expansion of the National Park Service in the 1930s* (Denver: Denver Service Center, National Park Service, 1983), 180, http://npshistory.com; Annie Robinson, "A 'Portrait of a Nation': The Role of the Historic American Buildings Survey in the Colonial Revival," in Wilson, Eyring, and Marotta, eds., *Re-Creating the American Past*, 104–5.

74. Robinson, "A 'Portrait of a Nation,'" 99. According to Wilton Claude Corkern Jr., a work relief program for architects had been discussed when creating Federal One, but no "Federal Architects Project" was created for three reasons: the American Institute of Architects did not pressure the WPA to set up a program; by mid-1935,

when the WPA was established, the number of unemployed architects had declined significantly; and HABS was already in place and demonstrated a successful record of employing professionals in the field. Corkern, "Architects, Preservationists, and the New Deal: The Historic American Buildings Survey, 1933–1942" (PhD diss., George Washington University, 1984), 126–27. For more on the beginnings of the HABS program, see Catherine C. Lavoie, "Architectural Plans and Visions: The Early HABS Program and Its Documentation of Vernacular Architecture," *Perspectives in Vernacular Architecture* 13, no. 2 (2006–7): 15–35.

75. Charles E. Peterson to the Director, U.S. Department of the Interior, Office of National Parks, Buildings, and Reservations, Washington, DC, November 13, 1933, reprinted in the *Journal of the Society of Architectural Historians* 16, no. 3 (October 1957): 29–31.

76. Ellen S. Woodward, "Allies in Aims: The Story of What America Is Doing to Preserve Its Historical Heritage," *National Historic Magazine*, December 1937, 1078, Box 15, Folder "290-B Historical Shrine Reconstruction and Preservation, Information Service, Primary File, 1936–42," Entry A1 678, RG 69, NACP.

77. Most HABS records are available digitally through the Library of Congress; see Historic American Buildings Survey/Historic American Engineering Record/Historic American Landscapes Survey Collection, Prints and Photographs Online Catalog, http://www.loc.gov. Corkern, "Architects, Preservationists, and the New Deal," 161–62; Lisa Pfueller Davidson and Martin J. Perschler, "The Historic American Buildings Survey during the New Deal Era: Documenting 'A Complete Resume of the Builders' Art,'" *CRM: The Journal of Heritage Stewardship* 1, no. 1 (Fall 2003): 49–73, https://www.nps.gov.

78. Robinson, "A 'Portrait of a Nation,'" 105, 114.

79. Michael Wallace, "Visiting the Past: History Museums in the United States," in *Presenting the Past: Essays on History and the Public*, ed. Susan Porter Benson, Stephen Brier, and Roy Rosenzweig (Philadelphia: Temple University Press, 1986), 149.

80. Meringolo, *Museums, Monuments, and National Parks*, 112.

81. Examples of CCC preservation work include the Carmel Mission in Carmel, California; Montezuma Castle National Monument in Camp Verde, Arizona; Fort Churchill in Lyon, Nevada; Mission Tejas State Park in Grapeland, Texas; and Saratoga National Historical Park in Stillwater, New York. Unrau and Williss, *Administrative History*, appendix 5: "National Park Service and CCC." For excellent histories of the Civilian Conservation Corps, see Maher, *Nature's New Deal*, and John C. Paige, *The Civilian Conservation Corps and the National Park Service, 1933–1942: An Administrative History* (Washington, DC: National Park Service, Department of the Interior, 1985).

82. West, *Domesticating History*, 130.

83. The CCC restored 3,980 historic structures according to the "Accomplishments of the Civilian Conservation Corps," cited in Leighninger, *Long-Range Public Investment*, 27, 30.

84. This division also employed Navajo workers to restore ancestral Pueblo ruins at Chaco Canyon, New Mexico, in cooperation with the NPS and Navajo Indian Service. Examples are drawn from the Living New Deal digital project administered by the Department of Geography, University of California, Berkeley, https://livingnewdeal.org. For a discussion of the challenges and benefits of the totem project, see McLerran, *A New Deal for Native Art*, 205–23.

85. See Edwin A. Lyon, *A New Deal for Southeastern Archaeology* (Tuscaloosa: University of Alabama Press, 1996); Paul Fagette, *Digging for Dollars: American Archaeology and the New Deal* (Albuquerque: University of New Mexico, 1996); and Bernard K. Means, ed., *Shovel Ready: Archaeology and Roosevelt's New Deal for America* (Tuscaloosa: University of Alabama Press, 2013).

86. Signed into law by Theodore Roosevelt, the Antiquities Act of 1906 was born out of congressional concerns over the loss and damage of historic and archaeological sites in the American Southwest due to activity by explorers and looters. Meringolo, *Museums, Monuments, and National Parks*, 44–48; Historic Sites Act of 1935, 49 Stat. 666; 16 U.S.C. § 461–67.

87. J. Thomas Schneider carried out much of the study in the eastern United States by touring historic sites; talking to NPS historians, preservationists, and staff at Colonial Williamsburg; and gathering data on European preservation practices and legislation. Unrau and Williss, *Administrative History*, 187–88; West, *Domesticating History*, 129–30.

88. West, *Domesticating History*, 130.

89. Quoted in Hosmer, *Preservation Comes of Age*, 1:573.

90. Franklin D. Roosevelt to Rene L. DeRouen and Robert F. Wagner, April 10, 1935, quoted in Unrau and Williss, *Administrative History*, 189. The letter was drafted by Verne Chatelain.

91. See the full Historic Sites Act of 1935 in Unrau and Williss, *Administrative History*, appendix 7: "Historic Sites Act," and Meringolo, *Museums, Monuments, and National Parks*, 123.

92. Some examples of restoration work completed under the new NPS Restoration Policy of 1937 and with the labor of New Deal work relief programs include Hopewell Village (now Hopewell Furnace National Historic Site) in Berks County, Pennsylvania (CCC and WPA); Fort Pulaski in Georgia (CCC, CWA, and Public Works Administration); and the Derby and Central Wharves at the Salem Maritime National Historic Site in Massachusetts (WPA and Public Works Administration). Unrau and Williss, *Administrative History*, 223.

93. Unrau and Williss, *Administrative History*, appendix 2: "Restoration Policies, May 19, 1937, Development of Restoration and Preservation Policies: 1935–1941."

94. In her study of historic preservation in Canada between 1920 and 1938, Shannon Ricketts found that the Canadian government looked to American efforts as models, particularly NPS-sponsored restorations and Colonial Williamsburg. Ricketts, "Cultural Selection and National Identity: Establishing Historic Sites in a National Framework, 1920–1939," *Public Historian* 18, no. 3 (July 1996): 33.

95. Calamia, "A New Deal for Historic Preservation," 56.
96. S. R. Winters, "Movement to Save the Nation's Heritage Has Spread Throughout the Country," *New York Times*, July 2, 1939.
97. For an excellent overview of the WPA bureaucratic operations, see Leighninger, *Long-Range Public Investment*, 55–79.
98. For more complicated and large-scale projects, the WPA often consulted with other appropriate agencies including the U.S. Army Corps of Engineers, Public Health Service, Department of Agriculture, Public Roads Administration, and Civil Aeronautics Administration. Smith, *Building New Deal Liberalism*, 113.
99. Jeff Hill, *The WPA: Putting America to Work* (Detroit, MI: Omnigraphics, 2014), 50–52.
100. The National Youth Administration initially assisted young people aged sixteen, but the age increased to eighteen in 1936. An excellent discussion of the NYA as Roosevelt's "model army" of modernity can be found in Kenneth J. Bindas, *Modernity and the Great Depression: The Transformation of American Society, 1930–1941* (Lawrence: University Press of Kansas, 2017), 68–81.
101. Carol A. Weisenberger, *Dollars and Dreams: The National Youth Administration in Texas* (New York: Peter Lang, 1994), 70.
102. Besides La Villita, an example of the broad construction and cultural work the NYA often performed within one project is the Daniel Boone Homestead in Birdsboro, Pennsylvania. Between 1938 and 1942, over one hundred youth workers built recreational structures, developed trails, planted trees, excavated a lake, constructed the Wayside Lodge, and restored the boyhood home of the American frontiersman. Calamia, "A New Deal for Historic Preservation," 124–34.
103. Carroll Scogin-Brincefield, "'The Yield on This Investment Should Be High': The National Youth Administration in Texas," in *Conflict and Cooperation: Reflections on The New Deal in Texas*, ed. Milton S. Jordan and George M. Cooper (Nacogdoches, TX: Stephen F. Austin State University Press, 2019), 67.
104. Griswold, *American Guides*, 3, 11, 91.
105. Hill, *The WPA*, 50–52.
106. Charles I. Glicksberg, "The Federal Writers' Project," *South Atlantic Quarterly* 37, no. 2 (April 1938): 158, 162.

Chapter 2

1. The title of the establishment is listed in sources as either the Planter's Hotel, the Planters' Hotel, or the Planters Hotel. For consistency, I have adopted the name most used by the architects during the WPA restoration: Planters' Hotel. Epigraph source is Robert Armstrong Andrews, "Dock Street Theatre," in "Work News—South Carolina," February 1937, Box 51, Folder "January 1937, Press Information and Publicity Materials, 1936–1942," (Entry A1 734), RG 69, NACP.
2. For discussion of twentieth-century efforts to preserve cultural heritage in Charleston, see Yuhl, *A Golden Haze of Memory*; Robert R. Weyeneth, *Historic Preservation for a Living City: Historic Charleston Foundation, 1947–1997* (Columbia:

University of South Carolina Press, 2000); and W. Fitzhugh Brundage, "Exhibiting Southernness in a New Century," in Brundage, *The Southern Past: A Clash of Race and Memory* (Cambridge, MA: Belknap Press of Harvard University Press, 2005), 183–226. Similar efforts to cope with the loss of social and political power and shape historical memory can be seen in the Natchez Garden Club's organization of the Pilgrimage around grand antebellum homes in the 1930s in Natchez, Mississippi. See Falck, *Remembering Dixie*, 153–210.

3. Many of the leading figures of the Charleston Renaissance, such as artists Elizabeth O'Neill Verner and Alfred Hutty, regarded the city's architecture as its most distinguishing feature and made Charleston's built environment the focus of their work. Novelist and poet Josephine Pinckney used the Dock Street Theatre restoration specifically to symbolize the New Deal era in a play written for the Carolina Art Association and intended to be performed at Middleton Place, an eighteenth-century rice plantation and major tourist attraction. The 1946 project was never completed. Barbara L. Bellows, *A Talent for Living: Josephine Pinckney and the Charleston Literary Tradition* (Baton Rouge: Louisiana State University Press, 2006), 206. For further discussion of the Charleston Renaissance, see James M. Hutchisson and Harlan Greene, eds., *Renaissance in Charleston: Art and Life in the Carolina Low Country, 1900–1940* (Athens: University of Georgia Press, 2003).

4. "1736–1937: In Commemoration and Rededication of the Dock Street Theatre, Charleston, S.C." (Charleston, SC: City of Charleston, 1937), 792 Se8, South Caroliniana Library, University of South Carolina, Columbia (hereafter SCL).

5. The Dock Street Theatre was built on a lot listed as no. 113 and registered to Nicholas Barlicorn on a 1693 map of Charleston. It was built forty-nine feet west of Church Street and seventy feet in front on Queen Street. Emmett Robinson, "A Guide to the Dock Street Theatre and Brief Resume of the Theatres in Charleston, S.C. from 1730" (Charleston: The Footlight Players, 1963), 4, Lowcountry Tourism Collection, Lowcountry Digital Library, College of Charleston Libraries, https://lcdl.library .cofc.edu.

6. During the WPA's restoration of the site, there was an unsuccessful "movement on foot to change the name of Queen Street back to Dock Street. . . . The old name is inseparably associated with the theatre." Donald Corley, "Restoration Footlights," n.d., Box 2530, Folder 651.1, S.C. Dock Street Theatre (hereafter Folder 651.1), Administrative and Operational Correspondence Relating to South Carolina, 1935-44 (hereafter Entry PC-37 12-45), RG 69, NACP; Thomas Cooper, ed., *The Statues at Large of South Carolina*, vol. 3, *1716–1752* (Columbia, SC: A. S. Johnston, 1838), 403; Julia Curtis, "A Note on Henry Holt," *South Carolina Historical Magazine* 79, no. 1 (January 1978): 1n1, 3–4.

7. Martha Zierden et al., "The Dock Street Theatre: Archaeological Discovery and Exploration," *Charleston Museum Archaeological Contributions* 42 (Charleston: Charleston Museum, 2009), 45–49. For discussion of early theaters in Charleston, see Hugh F. Rankin, *The Theater in Colonial America* (Chapel Hill: University of North Carolina Press, 1960); James H. Dormon Jr., *Theater in the Ante Bellum*

South, 1815–1861 (Chapel Hill: University of North Carolina Press, 1967); Charles S. Watson, *Antebellum Charleston Dramatists* (University: University of Alabama Press, 1976); and Watson, *The History of Southern Drama* (Lexington: University Press of Kentucky, 1997).

8. When Englishman John Lambert visited Charleston in 1809, he stayed at the Calders' first Planters' Hotel (before its relocation to the former site of the Dock Street Theatre). Of the establishment he wrote, "There are four or five hotels and coffee-houses in Charleston; but, except the Planters' hotel in Meeting-street, there is not one superior to an English public-house. The accommodations at the Planters' hotel are respectable, and the price about twelve dollars a-week." Jennie Holton Fant, ed., *The Travelers' Charleston: Accounts of Charleston and Lowcountry, South Carolina, 1666–1861* (Columbia: University of South Carolina Press, 2016), 103n, 104n. While the Calders were away in Europe, an innkeeper named Orran Byrd presided over the establishment. He added nine rooms and expanded the stables in 1816. "30. Dock Street Theater, 1736," Alfred O. Halsey Map 1949, Preservation Society of Charleston, http://www.halseymap.com.

9. Eola Willis donated her theatrical collections to the Dock Street Theatre project, requested to be "curator," and wanted complimentary use of the auditorium to deliver historical talks about theater. She thought these requests reasonable considering, she wrote WPA director Harry Hopkins, "that if I hadn't made years of research to discover name and site, there *wouldn't have been any second Dock Street Theatre.*" Eola Willis to Harry L. Hopkins, July 10, 1935, Box 2530, Folder 651.1, Entry PC-37 12-45, RG 69, NACP.

10. Eola Willis, *The Charleston Stage in the XVIII Century: With Social Settings of the Time* (Columbia, SC: State Company, 1924), 22; Sidney R. Bland, "Women and World's Fairs: The Charleston Story," *South Carolina Historical Magazine* 94, no. 3 (July 1993): 166–84.

11. A German visitor to the city in 1839 commented on the hostelry's popularity and success, noting that "because of many guests, we had to be satisfied with a tiny room. The [former] proprietor netted $30,000 after a mere fourteen years of hotel keeping." Frederic Trautmann, "South Carolina through a German's Eyes: The Travels of Clara von Gerstner, 1839," *South Carolina Historical Magazine* 85, no. 3 (July 1984): 221.

12. "30. Dock Street Theater, 1736."

13. A special correspondent to the *New York Times* in 1861 met with volunteers and officers in the hostelry. "From Charleston: A Slight Difference of Opinion about That Firing," *New York Times*, March 14, 1861; "Affairs at Fort Sumpter—Messengers Going To and Fro," *New York Times*, March 26, 1861.

14. Laylon Wayne Jordan, "'The Method of Modern Charity': The Associated Charities Society of Charleston, 1888–1920," *South Carolina Historical Magazine* 88, no. 1 (January 1987): 44.

15. Ellen Noonan, *The Strange Career of Porgy and Bess: Race, Culture, and America's Most Famous Opera* (Chapel Hill: University of North Carolina Press, 2012), 56, 72, 129.

16. Works that address the changing sociopolitical dynamic between white and Black residents in Charleston following the Civil War include Wilbert L. Jenkins, *Seizing the New Day: African Americans in Post–Civil War Charleston* (Bloomington: Indiana University Press, 1998); Katie Ann Stojsavljevic, "Housing and Living Patterns among Charleston's Free People of Color in Wraggborough, 1796–1877" (MA thesis, Clemson University, 2007); Bernard Edward Powers Jr., *Black Charlestonians: A Social History, 1822–1885* (Fayetteville: University of Arkansas Press, 1994); and Jeffrey Strickland, "How the Germans Became White Southerners: German Immigrants and African Americans in Charleston, South Carolina, 1860–1880," *Journal of American Ethnic History* 28, no. 1 (Fall 2008): 52–69.

17. T. R. Waring Jr., "A Cradle of America's Theater Will Harbor the Drama Again," *New York Herald Tribune*, December 12, 1937.

18. The development of modern waterfront facilities began in 1912 farther up the Charleston Neck, thereby avoiding altering the historic built environment of the peninsula despite significant economic change. Don H. Doyle, *New Men, New Cities, New South: Atlanta, Nashville, Charleston, Mobile, 1860–1910* (Chapel Hill: University of North Carolina Press, 1990), 55–61, 162, 232.

19. Jack Irby Hayes Jr., *South Carolina and the New Deal* (Columbia: University of South Carolina Press, 2001), 3.

20. The Charleston Art Commission was made up of the mayor and representatives of the Carolina Art Association, Charleston Library Society, Charleston Museum, South Carolina Historical Society, and three citizens, whose duty was to review "matters affecting the aesthetic and historic interests of the city" brought before the city council. Doyle, *New Men, New Cities, New South*, 226–28.

21. Some works that address Progressive-era concepts of race, cleanliness, and social order include Noralee Frankel and Nancy S. Dye, eds., *Gender, Class, Race, and Reform in the Progressive Era* (Lexington: University Press of Kentucky, 1991); Daniel Eli Burnstein, *Next to Godliness: Confronting Dirt and Despair in Progressive Era New York City* (Urbana: University of Illinois Press, 2006); and Kristen R. Egan, "Conservation and Cleanliness: Racial and Environmental Purity in Ellen Richards and Charlotte Perkins Gilman," *Women's Studies Quarterly* 39, nos. 3–4 (Fall–Winter 2011): 77–92.

22. Doyle, *New Men, New Cities, New South*, 228–38.

23. Hayes, *South Carolina and the New Deal*, 41–43; Sidney R. Bland, *Preserving Charleston's Past, Shaping Its Future: The Life and Times of Susan Pringle Frost*, 2nd ed. (1994; Columbia: University of South Carolina Press, 1999), 71; Stephanie E. Yuhl, "Rich and Tender Remembering: Elite White Women and an Aesthetic Sense of Place in Charleston, 1920s and 1930s," in Brundage, ed., *Where These Memories Grow*, 227–48.

24. While Frost and the SPOD were not actively involved in the Dock Street Theatre's restoration in the 1930s, architect Albert Simons acknowledged they deserved "much credit for safeguarding the old Planters' Hotel buildings" when it was under threat of demolition in 1918. Albert Simons, "Dock St. Theater, Planters Hotel Add

to City's Architectural Wealth," *Charleston News and Courier*, November 20, 1937; Hayes, *South Carolina and the New Deal*, 42.

25. Yuhl, *A Golden Haze of Memory*, 166–67; Noonan, *The Strange Career of Porgy and Bess*, 129–30.

26. Yuhl, *A Golden Haze of Memory*, 45, 48–49; Noonan, *The Strange Career of Porgy and Bess*, 133.

27. Kieran W. Taylor, "Chronology," in *Charleston and the Great Depression: A Documentary History, 1929–1941*, ed. Kieran W. Taylor (Columbia: University of South Carolina Press, 2018), xv. For the controversy over federal housing projects, which resulted in the construction of Cooper River Court for Blacks and Meeting Street Manor for whites, see Hayes, *South Carolina and the New Deal*, 74, and T. Robert Hart, "The Lowcountry Landscape: Politics, Preservation, and the Santee-Cooper Project," *Environmental History* 18, no. 1 (January 2013): 151.

28. Regina Bures and William Kanapaux, "Historical Regimes and Social Indicators of Resilience in an Urban System: The Case of Charleston, South Carolina," *Ecology and Society* 16, no. 4 (December 2011): 16.

29. Doyle, *New Men, New Cities, New South*, 301–2.

30. For discussion of local resistance, particularly from landowners, to the Santee-Cooper hydroelectric project, see Hart, "The Lowcountry Landscape."

31. Hayes, *South Carolina and the New Deal*, 42.

32. Writers' Program of the Work Projects Administration in the State of South Carolina, *South Carolina: A Guide to the Palmetto State* (New York: Oxford University Press, 1941), 187.

33. Elizabeth O'Neill Verner, "Preserving Our Own," *Charleston News and Courier*, January 26, 1938.

34. For discussion of the development of gentrification as it intersected with historical tourism in twentieth-century Charleston, see Yuhl, *A Golden Haze of Memory*; Weyeneth, *Historic Preservation for a Living City*; and Regina Bures, "Historic Preservation, Gentrification, and Tourism: The Transformation of Charleston, South Carolina," in *Critical Perspectives on Urban Redevelopment*, ed. K. F. Gotham (New York: Elsevier, 2001), 195–210.

35. Brittany V. Lavelle Tulla, "Huger, Cleland Kinloch and Burnet R. Maybank House," National Register of Historic Places Nomination Form (Washington, DC: U.S. Department of the Interior, National Park Service, 2015), section 8, p. 16.

36. E. M. Collison, "Project Reaches Fruition 3 Years after Inception," *Charleston Evening Post*, November 26, 1937; Simons, "Dock St. Theater, Planters Hotel Add to City's Architectural Wealth."

37. As evidence of their close relationship, Hopkins and his wife spent Thanksgiving with the Maybanks in 1936, and the mayor was a pallbearer at Hopkins's wife's funeral the following year. "Mr. Hopkins Off after Visit Here," *Charleston News and Courier*, November 27, 1936; "Hopkins Visiting Here," *Charleston News and Courier*, October 13, 1937.

38. Simons, "Dock St. Theater, Planters Hotel Add to City's Architectural Wealth."

39. Federal funds covered 90 percent of FERA relief spending in the South compared to 62 percent nationwide. Anthony J. Badger, *New Deal/New South: An Anthony J. Badger Reader* (Fayetteville: University of Arkansas Press, 2007), 36. For an overview of the New Deal's focus on developing the southern region prior to 1938, see Bruce J. Schulman, *From Cotton Belt to Sunbelt: Federal Policy, Economic Development, and the Transformation of the South, 1938–1980* (Durham, NC: Duke University Press, 1994), 3–38.

40. Maybank remained popular during his seven years as mayor from 1931 to 1938. His strategic backing of popular New Deal projects brought over $36 million of federal aid to the Lowcountry between 1933 and 1936 and earned him political support, leading to his victory in the South Carolina gubernatorial election of 1938. Maybank later was elected to the U.S. Senate in 1941, a position he held until his death in 1954. For further discussion of Maybank's control over New Deal measures in Charleston, see Alan Brinkley, "The New Deal and Southern Politics," in *The New Deal and the South*, ed. James C. Cobb and Michael V. Namorato (Jackson: University Press of Mississippi, 1984), 97–115, and Marvin L. Cann, "Burnet Maybank and Charleston Politics in the New Deal Era," in *Proceedings of the South Carolina Historical Association* (Columbia: South Carolina Historical Association, 1970), 39–48.

41. In an oral interview from the 1990s, Charlestonian Mary Ann Pearlstine recalled a family story in which her grandfather Hyman Pearlstine won the Planters' Hotel in a card game and offered to donate the property to the city of Charleston during the Depression if it would forgive the taxes he owed on the property. Various accounts of the restoration of the Dock Street Theatre describe the transfer of the property from Milton Alfred Pearlstine to the city as a philanthropic gift, but a copy of the financial status of the Dock Street Theatre WPA project of March 31, 1937, lists the sponsor's (the city's) contribution as $7,500 for "the purchase of the old property." "Jewish Heritage Collection: Oral History Interview with Mary Ann Pearlstine Aberman and Edward Aberman," interview by Dale Rosengarten and Barbara Karesh Stender, September 23, 1999, Mss 1035–222, Jewish Heritage Collection, Lowcountry Digital Library, College of Charleston Libraries, https://lcdl.library.cofc.edu; Colonel Francis C. Harrington to C. K. Yeager, Memorandum, April 3, 1937, Box 2530, Folder 651.1, Entry PC-37 12-45, RG 69, NACP.

42. "Planters' Hotel to Be Restored," *Charleston News and Courier*, February 5, 1935.

43. "Planters' Hotel Work Is Started," *Charleston News and Courier*, February 13, 1935; "Hopkins Praises Grice's Job Here," *Charleston News and Courier*, March 23, 1935.

44. "Planters' Hotel Is Being Stripped," *Charleston News and Courier*, June 19, 1935; "For Release Friday, June 7, 1935. Construction Starts on Rebuilding America's First Theater. No. 1212," Box 2530, Folder 651.1, Entry PC-37 12-45, RG 69, NACP.

45. Douglas Ellington to Jacob Baker, Memorandum, July 9, 1935, Folder 651.1, Box 2530, Folder 651.1, Entry PC-37 12-45, RG 69, NACP.

46. Workers also unearthed the remains of two cisterns, one in the central portion of the old hotel and another behind the building, which proved the existence of

a residence at some point in time at the back of the hotel. The rotten timber was carted from the theater site to the wood yard on East Bay Street, where it was sawed into firewood for the poor. "Planters' Hotel Work Will Start," *Charleston News and Courier*, May 4, 1935; "Planters' Hotel Project to Begin," *Charleston News and Courier*, May 7, 1935; "FERA Head Visits City," *Charleston News and Courier*, May 9, 1935; "Work on Theater to Start in Week," *Charleston News and Courier*, May 25, 1935; "Planters' Hotel is Being Stripped"; "Planters' Hotel Debris is Sifted," *Charleston News and Courier*, July 5, 1935; "Relics Excavated in Hotel Project," *Charleston News and Courier*, August 8, 1935. For more recent archaeological discoveries of the site, see Zierden et al., "The Dock Street Theatre."

47. Eola Willis to Douglas Ellington, December 4, 1935; Lester E. Lang to Helen Smith Whaley, September 16, 1935, Box 2530, Folder 651.1, Entry PC-37 12-45, RG 69, NACP.

48. The Footlight Players' president also wrote to Lawrence M. Pinckney, WPA state administrator, with the same interest on September 10, 1935. J. P. Frost to Hallie Flanagan, September 9, 1935; Flanagan to Frost, September 12, 1935; Frost to Flanagan, September 14, 1935; Frost to Belford Forrest, October 22, 1935, Container 195, Box 21, Folder 25-192-2 (1937), Carolina Art Association: Dock Street Theatre 1937, Carolina Art Association Correspondence, 1933–54, Dock Street Theatre Collection, 1933–58 1177.00 (hereafter DST), South Carolina Historical Society, Charleston (hereafter SCHS).

49. Hallie Flanagan to J. P. Frost, October 21, 1935; Flanagan to Daniel Reed, November 27, 1935, Box 2530, Folder 651.1, Entry PC-37 12-45, RG 69, NACP.

50. "Work on Theater to Start in Week."

51. In July 1935, Ellington resigned from the FERA to begin work with the Resettlement Administration, where he became the principal architect of Greenbelt, Maryland, the first planned community built by the federal government. The assistant administrator of the FERA wrote Mayor Maybank to assure him that Ellington "shall give as much time as necessary for general attention to the Charleston project." Ellington remained involved with the Dock Street Theatre project throughout its entirety. "Ellington Resign Government Post," *Charleston News and Courier*, July 18, 1935; "Start of Theatre Project Marked," *Charleston News and Courier*, August 15, 1935.

52. "Charlestonian Helps Revive Interest in American Architecture," *Charleston News and Courier*, December 25, 1938; "American Institute Gives Lapham Honor," *Charleston News and Courier*, May 30, 1937; Ralph Muldrow, "Simons and Lapham," South Carolina Encyclopedia, http://www.scencyclopedia.org; Hosmer, *Preservation Comes of Age*, 1:241.

53. Simons and Lapham edited published works of architectural history throughout their careers, including the coedited *The Octagon Library of Early American Architecture*, vol. 1, *Charleston, South Carolina* (New York: American Institute of Architects, 1927), and *Plantations of the Carolina Low Country* (Charleston, SC: Carolina Art Association, 1939).

54. "Byrnes and Hopkins Speak at Luncheon," *Charleston Evening Post*, March 22, 1935; "Hopkins to Dedicate Historic Theater Restored by WPA Workers," WPA Announcement No. 4–1615, November 21, 1937, Box 11, Folder "H.S.—SC—Dock Street Theatre & Planters' Hotel—Textual, Records Concerning the Restoration of Historic Shrines," Entry A1 764, RG 69, NACP.

55. "Restoration style" refers to architecture built in England during the period from the restoration of the Stuart monarchy in 1660 to the end of Charles II's reign in the 1680s. Erin Shaw, "Preservation Prologue: Albert Simon's Adaptive Reuse of the Planters' Hotel as the Dock Street Theatre" (seminar paper, University of South Carolina, Columbia, 1998), 7, Manuscripts, Schulz, C.B., SCL.

56. "Byrnes and Hopkins Speak at Luncheon."

57. "For Release Friday, June 7, 1935," Box 2530, Folder 651.1, Entry PC-37 12-45, RG 69, NACP.

58. Kenneth Chorley and Harold Shurtleff sent Douglas Ellington "documentary evidence which substantiated the statement that the first theater in America was built in Williamsburg." Chorley wrote a letter to Harold L. Ickes, which was then forwarded to Harry Hopkins, expressing Chorley's concerns about the claim of the Dock Street Theatre being the nation's first public theater. Chorley wrote, "We have no desire to enter into a quarrel with Charleston . . . but it does seem, I think, unfortunate that Federal funds should be used for a project the basis of which is unfounded." Kenneth Chorley to Harold L. Ickes, October 25, 1935; Harry L. Hopkins to Chorley, November 4, 1935, Chorley to Hopkins, November 18, 1935, Box 2530, Folder 651.1, Entry PC-37 12-45, RG 69, NACP; Kammen, *Mystic Chords of Memory*, 476.

59. "Theatre and Hotel of 1736 Restored," *New York Times*, November 21, 1937.

60. "First U.S. Theatre Restored with Help of American Seating Co.," *Motion Picture Herald* (Chicago), 130, no. 4 (February 5, 1938): 3.

61. Thomas P. Lesesne, "Theatre Serves Past and Present," *Charleston News and Courier*, November 27, 1937.

62. "Historic Theater, 200 Years Old, Reopened by U.S.," *Atlanta Constitution*, November 27, 1937.

63. Albert Simons to Emmett Robinson, March 20, 1936, Container 195, Box 21/195, Folder 25-192-2, DST, SCHS.

64. Simons, "Dock St. Theater, Planters Hotel Add to City's Architectural Wealth."

65. "Planters' Hotel Project to Begin"; "Planters' Hotel Details Mapped," *Charleston News and Courier*, May 31, 1935; "Fine Restaurant for Dock Street Theatre," *Charleston News and Courier*, June 27, 1937. In addition to the auditorium, the architects designed an entrance lobby, dining room and dining cloister, open courtyard, smoking room, bar, dressing rooms, offices, and committee rooms. They also included a green room, traditionally a retiring place for actors but which would also host lectures, rehearsals, and small concerts. Douglas Ellington, Samuel Lapham, and Albert Simons, "Charleston Opens Historic Playhouse with Historic Play," *Architectural Record* 83, no. 1 (January 1938): 20–25.

66. Rowena Wilson Tobias, "Charleston Preparing to Reopen Oldest Theater with Historic Play," *State* (Columbia, SC), November 14, 1937.

67. Simons, "Dock St. Theater, Planters Hotel Add to City's Architectural Wealth."

68. "Restored Ancient Theater," *State*, January 24, 1937.

69. Shaw, "Preservation Prologue," 9; Corley, "Restoration Footlights."

70. Ellington, Lapham, and Simons, "Charleston Opens Historic Playhouse with Historic Play," 21.

71. Ellington, Lapham, and Simons, "Charleston Opens Historic Playhouse with Historic Play," 21.

72. Calgar Aydin, "The Potential of Virtual Heritage Reconstruction in Lost Ansonborough" (MSc thesis, Clemson University, 2012), 17–23.

73. Weyeneth, *Historic Preservation for a Living City*, 6.

74. When the city council approved the demolition of the building to make way for the College of Charleston's new gymnasium in 1938, it hired Albert Simons to design the new gym. He utilized the old mansion's masonry walls and the iron fence delineating the property's perimeter in its construction. "Old High School Is Put on Block," *Charleston News and Courier*, August 15, 1935.

75. Douglas D. Ellington to Samuel Lapham, July 18, 1935, Box 2530, Folder 651.1, Entry PC-37 12-45, RG 69, NACP.

76. Douglas D. Ellington to Edward P. Grice Jr., July 19, 1935, Box 2530, Folder 651.1, Entry PC-37 12-45, RG 69, NACP.

77. Aydin, "The Potential of Virtual Heritage Reconstruction in Lost Ansonborough," 18–20; Douglas Ellington to Jacob Baker, Memorandum, July 9, 1935.

78. Dubose Heyward, "Dock Street Theatre: Carolina Art Association Management," January 1938, 792, H49, 1938, Oversize, SCL.

79. Simons, "Dock Street Theater, Planters Hotel Add to City's Architectural Wealth."

80. Collison, "Project Reaches Fruition."

81. Brundage, *The Southern Past*, 208.

82. While the Dock Street Theatre harkens back to the colonial period, the Civil War generally loomed large in New Deal historical projects of the 1930s and 1940s. See Nina Silber, *This War Ain't Over: Fighting the Civil War in New Deal America* (Chapel Hill: University of North Carolina Press, 2018).

83. Heyward, "Dock Street Theatre." Ellen Noonan has described Heyward as "prominent" among Charleston preservationists as "their poet laureate and chief publicist." Noonan, *The Strange Career of Porgy and Bess*, 131.

84. Leora Auslander, "Beyond Words," *American Historical Review* 110, no. 4 (October 2005): 1020.

85. Carolina Art Association, "Untitled," n.d., 3–4, Container 195, Box 21, Folder 21-195-2 (1937), DST, SCHS.

86. The contract for stage equipment was awarded to Bruckner Mitchell Company of New York City. Douglas D. Ellington to C. E. Tompkins, February 5, 1937, Box 2530, Folder 651.1, Entry PC-37 12-45, RG 69, NACP.

87. Heyward, "Dock Street Theatre."

88. "National Theater Germ Seen Here," *Charleston News and Courier*, September 23, 1937.

89. Burnet R. Maybank to Douglas D. Ellington, February 12, 1936, Box 2530, Folder 651.1, Entry PC-37 12-45, RG 69, NACP.

90. Daniel Ravenel to Douglas D. Ellington, March 10, 1936, Box 2530, Folder 651.1, Entry PC-37 12-45, RG 69, NACP.

91. Samuel Lapham to Douglas D. Ellington, June 9, 1936, Box 2530, Folder 651.1, Entry PC-37 12-45, RG 69, NACP.

92. The theater, which had become a federal work relief project under the FERA in February 1935, was taken over by the WPA officially on November 11, 1936. "Dock Street Theatre Funds Awaited," *Charleston News and Courier*, October 10, 1936; "Dock Street Theatre Restoration Halted; WPA Coin Used Up," *Variety*, October 21, 1936; "President Allots Dock Street Fund," *Charleston News and Courier*, October 25, 1936; "Theater Project Gets Some Funds," *Charleston News and Courier*, November 12, 1936.

93. E. P. Grice Jr. to Lawrence M. Pinckney, March 20, 1937; Albert Simons to Pinckney, March 20, 1937, Box 2530, Folder 651.1, Entry PC-37 12-45, RG 69, NACP.

94. Albert Simons to Burnet R. Maybank, January 9, 1937, Box 2530, Folder 651.1, Entry PC-37 12-45, RG 69, NACP.

95. Harry Hopkins to Burnet R. Maybank, February 19, 1937, Box 2530, Folder 651.1, Entry PC-37 12-45, RG 69, NACP.

96. Albert Simons to Douglas D. Ellington, February 11, 1936, Box 2530, Folder 651.1, Entry PC-37 12-45, RG 69, NACP.

97. For example, in the summer of 1936, Maybank successfully entreated WPA deputy administrator Aubrey Williams for an additional $35,000 because labor expenses had been higher than expected. Burnet R. Maybank to Aubrey Williams, Telegraph, September 26, 1936. This and other funding requests are located in Box 2530, Folder 651.1, Entry PC-37 12-45, RG 69, NACP.

98. "Dock St. Theater Done by Spring," *Charleston News and Courier*, November 21, 1936; "Restored Ancient Theater."

99. The electrical work was done by the Southern Electric Company and the plumbing and heating by W. K. Prause, whose shop was on Charleston's King Street. Samuel Lapham Jr. to Douglas D. Ellington, November 10, 1935, Box 2530, Folder 651.1, Entry PC-37 12-45, RG 69, NACP.

100. Simons, "Dock St. Theater, Planters Hotel Add to City's Architectural Wealth."

101. "Hopkins to Dedicate Historic Theater Restored by WPA Workers."

102. R. P. Harriss, "Charleston Ponders What to Do with Its New WPA Playhouse," n.d., Box 11, Folder "H.S.—SC—Dock Street Theatre & Planters' Hotel—Textual," Entry A1 764, RG 69, NACP.

103. William Watts Ball became editor of the *News and Courier* in 1927 and began publicly criticizing New Deal projects, especially the Santee-Cooper hydroelectric dam, in its pages in 1933. Herbert Ravenel Sass, *Outspoken: 150 Years of The News and Courier* (Columbia: University of South Carolina Press, 1953). For more on his

conservative editorship, particularly surrounding issues of race, see Sid Bedingfield, *Newspaper Wars: Civil Rights and White Resistance in South Carolina, 1935–1965* (Urbana: University of Illinois Press, 2017), 83–107.

104. "Ever Been to Jail," *Charleston News and Courier*, March 29, 1937.

105. "$870,000 a Month," *Charleston News and Courier*, December 10, 1937.

106. Jane De Hart Mathews, *The Federal Theatre, 1935–1939: Plays, Relief, and Politics* (Princeton, NJ: Princeton University Press, 1967).

107. The union member mentioned an agreement Hallie Flanagan made with the International Alliance of Theatrical Stage Employees that only members of that organization could be employed to do stage work or build scenery on FTP projects. W. O. Duc to Hallie Flanagan, September 8, 1936, Box 2530, Folder 651.1, Entry PC-37 12-45, RG 69, NACP.

108. The other two projects named were "the Greenville textile museum" and "the Columbia art gallery." "WPA Hopes to Keep Artists at Home, National Head Says," *Charleston News and Courier*, April 9, 1936.

109. "Art Association May Run Theater," *Charleston News and Courier*, June 19, 1937; Harold A. Mouzon and Robert N. S. Whitelaw, "Carolina Art Association to Ways and Means Committee, City Council, City of Charleston, SC," February 4, 1948, 1253.00, Folder 26-35-11, Albert Simons Papers, 1864–1979, SCHS; "Playhouse Contract Signed by Maybank," *Charleston News and Courier*, November 11, 1937.

110. Robert N. S. Whitelaw, "Whitelaw Tells of Future Plans," *Charleston News and Courier*, December 5, 1937; Simons to Robinson, March 20, 1936.

111. Untitled, Spring 1938, Container 195, Box 21/195, Folder 25-192-2, DST, SCHS.

112. Jacob Baker to Lawrence M. Pinckney, April 21, 1936, Box 2530, Folder 651.1, Entry PC-37 12-45, RG 69, NACP.

113. Rhett also painted the English coat of arms over the proscenium arch surrounding the stage. Six additional backdrops were created for the comedy. "Charleston of 1838 Shown in Dock Street Scenery," *Charleston News and Courier*, November 24, 1937; Rowena Wilson Tobias, "New Dock Street Theater to Repeat Comedy Given More Than 200 Years Ago," *State*, November 25, 1937.

114. "Alicia Rhett Off for Film Test in North," *Charleston News and Courier*, March 14, 1937; Waring, "A Cradle of America's Theater Will Harbor the Drama Again."

115. "Curtains Are Being Hung," Box 11, Folder "H.S.—SC—Dock Street Theatre & Planters' Hotel—Textual," Entry A1 764, RG 69, NACP.

116. The state guidebook describes this plasterwork but does not name Smith as the artisan, instead referring to him as "an ex-slave" who was "the only artisan considered competent to take charge of the plasterwork in the enlarged clerestory." Simons, "Dock St. Theater, Planters Hotel Add to City's Architectural Wealth"; Writers' Program of the WPA, *South Carolina*, 117.

117. Waring, "A Cradle of America's Theater Will Harbor the Drama Again."

118. William Halsey and Corrie McCallum, interview by Liza Kirwin, October 27, 1986, Archives of American Art Oral History Program, Smithsonian Institution, https://www.aaa.si.edu.

119. The Academy of Music, a structure built in 1838 and converted into a theater in 1869, was demolished in 1937, the year William Halsey painted the courtyard fresco. The four panels selected were *The Enraged Musician, Modern Midnight Conversation,* and two prints from the series The Rake's Progress: *The Gaming House* and *The Madhouse.* "Halsey Frescoes Theater's Mural," *Charleston News and Courier,* October 3, 1937.

120. Three hundred tickets were distributed to patriotic and civic organizations across the nation, including the directors of the Metropolitan Opera Company and the New England Conservatory of Music; three hundred to city boards and commissions; two hundred to federal and state officials; fifty to the Charleston City Council members and guests; fifty to military units in the Charleston area; fifty to county officials; and fifty to colleges, newspapers, and dramatic critics. "Playhouse Debut Bids Are Issued," *Charleston News and Courier,* November 16, 1937.

121. First Lady Eleanor Roosevelt and WPA deputy administrator Aubrey C. Williams were also invited but unable to attend the opening. "Hopkins to Give Theater to City," *Charleston News and Courier,* November 11, 1937; "Dock Street Theatre Restored," *State,* November 24, 1937; "Formal Opening Begins Tonight at Dock St. Theater," *Charleston News and Courier,* November 26, 1937; Albert Simons to Douglas D. Ellington, February 11, 1936, Box 2530, Folder 651.1, Entry PC-37 12-45, RG 69, NACP.

122. "Hopkins to Dedicate Historic Theater Restored by WPA Workers."

123. Letters from the Associated Press, November 11, 1937, *Washington Post,* December 24, 1937, and *Time,* October 18, 1937, Box 21/195, Folder 21-195-2, DST, SCHS; "Dock Street Theatre Restored."

124. "Dock Street Theater Ushers at Formal Opening to Be Cadets and Collegians," *Charleston News and Courier,* November 19, 1937; "Souvenirs Will Be Presented at Opening of Dock Street Theater," *Charleston Evening Post,* November 26, 1937; R. M. Hitt Jr., "City's Culture Made Theater Gift Possible, Hopkins Says," *Charleston News and Courier,* November 27, 1937.

125. After five tornadoes struck the Charleston peninsula on September 29, 1938, Harry Hopkins directed the WPA to give $500,000 to restore landmarks hit by the storms, including the Dock Street Theatre, once again lending his support to protect Charleston's built heritage. "WPA Gives $500,000 to Fix Landmarks; Hopkins Offers Immediate Help to Rebuild Here," *Charleston News and Courier,* October 1, 1938; "Dock St. Season to Go Forward," *Charleston News and Courier,* October 1, 1938.

126. Hitt, "City's Culture Made Theater Gift Possible."

127. Hitt, "City's Culture Made Theater Gift Possible."

128. E. M. Collison, "Performances at Dock St. Theater to Be Repeated," *Charleston Evening Post,* November 27, 1937.

129. "Roosevelt Waits to See Maybank," *Charleston News and Courier,* November 29, 1937.

130. "Charleston Opens Theatre as in 1736," *New York Times,* November 27, 1937; "First U.S. Theatre Is Restored: Charleston Blue Bloods Give It Gala Opening," *Life,*

December 20, 1937, 49–51; Ellington, Lapham, and Simons, "Charleston Opens Historic Playhouse with Historic Play"; Daisy Mae Roberts, "The Dock Street Theatre Rebuilt—A Federal Project," *Scholastic*, January 15, 1938, Box 11, Folder "H.S.—SC—Dock Street Theatre & Planters' Hotel—Textual," Entry A1 764, RG 69, NACP.

131. "Accomplishment of WPA Detailed," *Charleston News and Courier*, October 13, 1937.

132. The Dock Street Theatre was the only community theater to receive a grant from the foundation, and only two other grants promoting drama were awarded overall (to the University of North Carolina at Chapel Hill and Western Reserve University). "$15,000 Rockefeller Grant Given Dock Street Theater," *Charleston News and Courier*, May 24, 1938; "Rockefeller Report Cites Dock Street," *Charleston News and Courier*, August 8, 1939; Hayes, *South Carolina and the New Deal*, 43.

133. Other organizations that utilized the Dock Street Theatre included the Charleston Garden Club, Thespian Players, Charleston Museum, Ashley Hall, City Federation of Women's Clubs, Society for the Preservation of Spirituals, Charleston Free Library, St. Michael's Protestant Episcopal Church, St. Paul's Protestant Episcopal Church, First Church of Christ Scientist, Society of Colonial Wars, and the local chapter of B'nai B'rith. "Dock St. Theater Has Big Season," *Charleston News and Courier*, April 9, 1939.

134. In March 1938, Hallie Flanagan traveled to Charleston to discuss ways in which the Federal Theatre Project might help in the further development of the theater. "Theater Policies to Be Discussed," *Charleston News and Courier*, January 22, 1938; "Experts Advise a Longer Lease," *Charleston News and Courier*, February 3, 1938; "WPA Aid Offered Dock St. Theater," *Charleston News and Courier*, March 31, 1938.

135. "Carolina Art Group Gets Rockefeller Aid," *New York Times*, November 9, 1939.

136. H. P. Lovecraft to Herman Charles Koenig, January 12, 1936, cited in Taylor, ed., *Charleston and the Great Depression*, 69–71.

137. "1736–1937: In Commemoration and Rededication of the Dock Street Theatre."

Chapter 3

1. "The Henry Whitfield House in Guilford, Connecticut" (Guilford, CT: Board of Trustees, 1957), Folder "Whitfield State Hist. Site," 974.62 G956hw, Connecticut Museum of Culture and History, Hartford (formerly the Connecticut Historical Society).

2. Figure 4 in Ralph D. Smith's work shows a map of Guilford, Connecticut, c. 1672, with locations of the other stone houses of prominent townspeople Jasper Stillwell, John Higginson, and Samuel Desborough. Smith, *The History of Guilford, Connecticut, from Its First Settlement in 1639* (Albany, NY: J. Munsell, 1877), 17; Robert Blair St. George, *Conversing by Signs: Poetics of Implications in Colonial New England Culture* (Chapel Hill: University of North Carolina Press, 1998), 27–28.

3. St. George, *Conversing by Signs*, 27.

4. When Henry Whitfield died in 1657, he left the Old Stone House to his wife, Dorothy (née Sheaffe). In 1659, she sold the property to Major Robert Thompson,

a prominent London merchant. Upon his death in 1694, Thompson left the income of his property to his wife, Dame Frances, but the title to the Old Stone House remained with his male descendants, sons Joseph, William, and Robert. The house remained in the Thompson family for more than a century until Wyllys Elliot purchased it in 1772. Just over two weeks later, he sold it to Joseph Pynchon, who may have been the first owner to reside in the house since the Whitfields. Shortly before the Declaration of Independence was signed in 1776, loyalist Pynchon sold the house to Jasper Griffing. Jasper's son Nathanial purchased the home from his father in 1800. Nathanial was a magistrate, ship owner, merchant, and Guilford's delegate in 1818 to the constitutional convention in Hartford. After he died on September 17, 1845, the Guilford property passed to his son Frederick H. Griffing. When the unmarried Frederick died seven years later in 1852, the Old Stone House become the property of his mother, Sarah Brown Griffing, who then bequeathed it to her only surviving child, daughter Mary, in 1865. The property then descended to Mary and Henry Ward Chittenden's daughter, Sarah Brown, who married Henry D. Cone. Thus, since 1776 the Old Stone House had descended through the Griffing, Chittenden, and Cone families, all related through marriage, and was usually occupied by tenants. "Owners" and "Residents," exhibit text panels at the Henry Whitfield State Museum, Guilford, Connecticut (hereafter HWSM); "Old Stone House Opened Today," *Hartford [CT] Courant*, September 21, 1904.

5. Sources cited in Jan Cunningham, "Whitfield, Henry, House," National Historic Landmark Nomination Form (Washington, DC: U.S. Department of the Interior, National Park Service, 1997), 8.

6. Kammen, *Mystic Chords of Memory*, 206–14.

7. Veneration of the Puritans was not completely dead in the 1920s; many held President Calvin Coolidge as a living symbol of New England Puritanism. Kammen, *Mystic Chords of Memory*, 387–406, 378 (quotation); Karal Ann Marling, "Of Cherry Trees and Ladies' Teas: Grant Wood Looks at Colonial America," in Axelrod, ed., *The Colonial Revival in America*, 295. For more on the use of the Puritan tradition in the 1920s and 1930s, see Marling, *George Washington Slept Here*, 267–72; Frederick J. Hoffman, "Philistine and Puritan in the 1920s: An Example of the Misuse of the American Past," *American Quarterly* 1, no. 3 (Autumn 1949): 247–63; and Jan C. Dawson, *The Unusable Past: America's Puritan Tradition, 1830–1930* (Chico, CA: Scholars Press, 1984).

8. Marling, "Of Cherry Trees and Ladies' Teas," 295; Van Wyck Brooks, *The Flowering of New England* (New York: E. P. Dutton, 1936).

9. Susman, *Culture as History*, 41–42.

10. John W. Jeffries, *Testing the Roosevelt Coalition: Connecticut Society and Politics in the Era of World War II* (Knoxville: University of Tennessee Press, 1979), 3–4, 14.

11. Patterson, *The New Deal and the States*, 158.

12. Jeffries, *Testing the Roosevelt Coalition*, 25, 35.

13. Robert L. Woodbury, "Wilbur Cross: New Deal Ambassador to a Yankee Culture," *New England Quarterly* 41, no. 3 (September 1968): 323.

14. Wilbur L. Cross, *Connecticut Yankee: An Autobiography* (New Haven, CT: Yale University Press, 1943); Woodbury, "Wilbur Cross," 324.

15. Cecelia Bucki, *Bridgeport's Socialist New Deal, 1915–36* (Urbana: University of Illinois Press, 2001), 138, 181–82; Jeffries, *Testing the Roosevelt Coalition*, 23–24.

16. Woodbury, "Wilbur Cross," 327, 334, 331.

17. Lindgren, *Preserving Historic New England*, 7–8.

18. Butler, "Another City upon a Hill," 15–51. See also Gebhard, "The American Colonial Revival in the 1930s," 109–48.

19. Lindgren, *Preserving Historic New England*, 4–5.

20. Kammen, *Mystic Chords of Memory*, 254; Hosmer, *Presence of the Past*, 120–22, 166, 169, 301–2.

21. Cunningham, "Whitfield, Henry, House," 11–12; "Memorials of Our Country's Young Life," April 1875, in *Potter's American Monthly: An Illustrated Magazine of History, Literature, Science and Art, Vols. IV and V* (Philadelphia: John E. Potter, 1875), 271–72.

22. Cunningham, "Whitfield, Henry, House," 9, 12.

23. William B. Rhoads, "The Colonial Revival and the Americanization of Immigrants," in Axelrod, ed., *The Colonial Revival in America*, 341–61; Rhoads, "The Colonial Revival and American Nationalism," *Journal of the Society of Architectural Historians* 35, no. 4 (December 1976): 239–54; Hosmer, *Presence of the Past*, 301; Lydia Mattice Brandt, *First in the Homes of His Countrymen: George Washington's Mount Vernon in the American Imagination* (Charlottesville: University of Virginia Press, 2016), 113.

24. Anne C. Reilly, "A Local Commemoration of National Significance," in Bruggeman, ed., *Commemoration*, 41–49.

25. Marling, *George Washington Slept Here*, 131.

26. Hosmer, *Presence of the Past*, 120, 264. For discussion of nineteenth- and early twentieth-century women's preservation work, see Barbara J. Howe, "Women in the Nineteenth Century Preservation Movement," in Dubrow and Goodman, eds., *Restoring Women's History through Historic Preservation*, 17–36, and West, *Domesticating History*.

27. Jerry L. Cross, "Andrew Johnson Birthplace," Encyclopedia of North Carolina, 2006, https://www.ncpedia.org.

28. The date of the sale was September 28, 1900, and the property acquired from Sarah Brown Cone is described in a deed recorded in the Guilford Land Records, book 56, p. 419, August 20, 1900. The state appropriated $3,500 for the sale, the town of Guilford gave $3,000, residents donated between $500 and $1,000, and the Connecticut Society of Colonial Dames provided the remainder. "Old Stone House Opened Today"; "Making an Appropriation for a State Historical Museum," *Special Acts and Resolution of the General Assembly of the State of Connecticut* (Hartford, CT: Hartford Press, 1901), 515, State Archives, Connecticut State Library, Hartford (hereafter CSL).

29. "Old Stone House Opened Today"; "Report of the Committee Appointed by the Board of Trustees of the Henry Whitfield House, to Meet the Committee

on Appropriations of the Legislature of the State of Connecticut for the Years 1905–1906," Box 1, Folder 1, RG 024:001 Henry Whitfield House Records, 1768–1957 (hereafter HWHR), CSL.

30. "The Henry Whitfield House in Guilford Connecticut," 5; J. Frederick Kelly, "Restoration of the Henry Whitfield House, Guilford, Connecticut," *Old-Time New England* 29, no. 3 (January 1939): 77, 79.

31. Norman Isham quoted in Cunningham, "Whitfield, Henry, House," 13–14.

32. Karin Peterson, "Site Lines: The Making of the Henry Whitfield House Museum," Connecticut Explored, Winter 2010–11, 32–33, https://www.ctexplored.org.

33. Isham learned of the folding partitions through correspondence with the last owner of the house, Sarah Brown Cone, who recounted her grandmother's description of the property and changes her family made to the house. Sarah B. Cone to Norman M. Isham, May 8, 18, 1899, MS 442, Box 106, Folder 1511, Series V: Topical Files, 1684–1945, George Dudley Seymour Papers (hereafter GDS), Sterling Memorial Library, Yale University, New Haven, Connecticut (hereafter SML); Kelly, "Restoration of the Henry Whitfield House," 79.

34. "Report of the Committee Appointed by the Board of Trustees of the Henry Whitfield House."

35. Peterson, "Site Lines."

36. Peterson, "Site Lines."

37. Swigger, *"History Is Bunk,"* 21.

38. Samuel Hart, "Guilford among her Neighbors," from "Papers Read at the Formal Opening of the State Historical Museum in the Henry Whitfield House, September 21, 1904," in Connecticut Society of Colonial Dames of America, *Connecticut Colonial Houses: Henry Whitfield House, Guilford, 1640* (N.p.: N.p., n.d.), 728, C718gu, Accession No. 67518, CSL.

39. Lindgren, *Preserving Historic New England*, 3.

40. James M. Lindgren describes Norman Isham as highly regarded by SPNEA in Lindgren, *Preserving Historic New England*, 72.

41. Norman M. Isham, "The Work Done at the Henry Whitfield House," September 7, 1906, Box 2, Folder 55, MS 1156, Norman M. Isham Papers (hereafter NMI), SML; emphasis added.

42. George Dudley Seymour became an active supporter of preservation in Connecticut. In 1914, he purchased the Nathan Hale Homestead in Coventry, which he restored and then later gifted to the Connecticut Antiquarian and Landmarks Society. Governor Wilbur Cross served as the society's first advisory chair. Hosmer, *Preservation Comes of Age*, 1:170–71.

43. Smith's plans also appeared in his *History of Guilford*, published in 1877 after his death, and in Edward E. Atwater's *History of New Haven to the Present Time* (New York: W. W. Munsell, 1887). John Gorham Palfrey, *History of New England* (Boston: Little, Brown, 1859), measured drawings follow p. 16.

44. Norman Isham quoted in Cunningham, "Whitfield, Henry, House," 14.

45. Seymour threatened to retire from the Board of Trustees multiple times between 1921 and 1932 on account of the faulty restoration, but the governors of Connecticut,

first Rollin Woodruff and then John H. Trumball, refused to accept his resignation. See letters in Box 106, Folder 1513, GDS, SML, and Box 1, Folder 12, NMI, SML. George Dudley Seymour to Alfred Hammer, May 16, 1921; Seymour to the President and Trustees, Henry Whitfield House, June 20, 1921, Box 106, Folder 1513, GDS, SML.

46. George Dudley Seymour to Frederick Norton, April 11, 1921, Box 106, Folder 1513, GDS, SML.

47. George Dudley Seymour to Norman Isham, August 11, 1930, Box 106, Folder 1513, GDS, SML.

48. Kelly performed the duties of architect and restorationist for the New Haven Colony Historical Society from 1926 until his death in 1947.

49. J. Frederick Kelly, *Early Domestic Architecture of Connecticut* (New Haven, CT: Yale University Press, 1924), 29–30.

50. Patrick J. Mahoney, "J. Frederick Kelly: Constructing Connecticut's Architectural History," Connecticut History, March 21, 2021, https://connecticuthistory.org.

51. "The Aims of the Society for the Preservation of New England Antiquities," *Old-Time New England* 18, no. 2 (October 1947): vii.

52. Deborah G. Burness, "J. Frederick Kelly and the Colonial Revival in Connecticut" (BA thesis, Southern Connecticut State University, 1994), 34–35.

53. Walpole Society members were required to have "the social qualifications essential to the well-being of a group of like-minded persons." Early members included George Francis Dow, Henry du Pont, R. T. H. Halsey, and Norman Isham. Walpole Society Constitution, 1942, amended version, 8, cited in Burness, "J. Frederick Kelly and the Colonial Revival in Connecticut," 29–30; Kammen, *Mystic Chords of Memory*, 323–24. For an outstanding examination of Appleton and SPNEA within the context of Yankee progressivism, see Lindgren, *Preserving Historic New England*.

54. Seymour and Kelly began working together as early as 1926 when they successfully fought to save the Curtis-Rose House for the Gallery of Fine Arts at Yale University. Burness, "J. Frederick Kelly and the Colonial Revival in Connecticut," 29–32.

55. J. Frederick Kelly to Alfred A. Hammer, January 18, 1927, August 30, 1928, Box 1, Folder 5, HWHR, CSL.

56. Cunningham, "Whitfield, Henry, House," 13; Rhoads, "The Colonial Revival," 1:517–18; "Historical Activities," National Society of the Colonial Dames of America in Connecticut, https://nscda-ct.org.

57. J. Frederick Kelly to Alfred Hammer, May 10, 1930, Box 1, Folder 5, HWHR, CSL.

58. Hosmer, *Presence of the Past*, 300; Lindgren, *Preserving Historic New England*, 4, 9.

59. George Dudley Seymour to Norman Isham, August 11, 1930, Box 106, Folder 1513, GDS, SML.

60. According to Kelly, the removal indicated the size and position of some of the original window openings and revealed a lime whitewash that was probably mixed with skim milk to form a powerful adhesive agent. He also discovered traces of red paint, which indicated that the exterior walls were first whitewashed and then painted at a later date, perhaps to prevent more moisture from entering the masonry. Kelly, "Restoration of the Henry Whitfield House," 79–80.

61. In late January 1935, Republican congressman Ray C. Loper of Guilford introduced a bill requesting the appropriation in the General Assembly. "Whitfield House Vacancy Filled," *Shore Line Times* (Guilford, CT), February 7, 1935.

62. Members of the Restoration Committee, which was sometimes referred to as the Reconstruction Committee, included chair Evangeline Andrews of New Haven, secretary of the Board of Trustees and wife of Charles McLean Andrews, esteemed scholar of Connecticut history at Yale University; Annie B. Jennings, who served for many years as the vice regent for Connecticut of the Mount Vernon Ladies Association; Dr. Frederic T. Murlless Jr.; Dr. Walter R. Steiner; and Frederick Calvin Norton, who served as president of the Board of Trustees. "Whitfield House, Federal Grant," *Shore Line Times*, November 7, 1935.

63. J. Frederick Kelly, "The Henry Whitfield House in Guilford Connecticut with Plans for its Restoration" (New Haven, CT: Yale University Press, 1935), Folder "LIBRARY—Pamphlet House Booklet, 1935," HWSM.

64. Accountant of the State of Connecticut Tercentenary Commission to Evangeline Andrews, June 5, 1936, Box 1, Folder 7, HWHR, CSL.

65. Evangeline Andrews to Eleanor Little, April 20, 1935; Little to Andrews, May 15, 1935, Box 1, Folder 6, HWHR, CSL.

66. Eleanor Little to Roland Hooker, July 23, 1935, Box 1, Folder 6, HWHR, CSL.

67. Evangeline Andrews to the Works Progress Administration, September 9, 1935, Box 1, Folder 6, HWHR, CSL.

68. Typically, Kelly charged 15 percent of the total cost of a project, but he lowered his commission to 10 percent for the Whitfield House restoration. J. Frederick Kelly, Trustees Minutes, May 9, 1935, Box 1, Folder 3, HWHR, CSL.

69. Evangeline Andrews to William Sumner Appleton, November 9, 1935; Appleton to Andrews, November 11, December 3, 1935, Box 1, Folder 6, HWHR, CSL. James Lindgren wrote that back in 1914 Appleton expressed disinterest in the Whitfield House and doubts about Norman Isham's restoration, saying the house should have been left "openly and frankly a ruin." Lindgren, *Preserving Historic New England*, 147.

70. SPNEA had been instrumental in previous Guilford preservation endeavors, including the Colonial Dames' acquisition of the 1752 Joseph Webb House for the Dorothy Whitfield Society (the Guilford historical organization founded as an auxiliary of the Henry Whitfield House after its dedication as the state museum in 1897) and the Thomas Lee House, constructed around 1660, in nearby East Lyme for the East Lyme Historical Society. George Dudley Seymour, Annie Burr Jennings, J. Frederick Kelly, and William B. Goodwin—all with connections to the Henry Whitfield House—were members of the Connecticut Committee of SPNEA. William Sumner Appleton to Annie Jennings, November 13, 1935, Box 1, Folder 6, HWHR, CSL.

71. Evangeline Andrews to Frederick C. Norton, November 29, 1935, Box 1, Folder 6, HWHR, CSL.

72. Trustees Minutes, November 7, 1935, Box 1, Folder 3; Evangeline Andrews to William Sumner Appleton, November 4, 1935, Box 1, Folder 6, HWHR, CSL; Senator Matthew A. Daly to Governor Wilbur Cross, December 5, 1935, Box 401, Folder

"Whitfield House Project," RG 005:026, Series 2: Subject Files, 1931–39, Subseries 1935–37, Veterans Home Commission, Wilbur L. Cross Records (hereafter WLC), CSL.

73. The Henry Whitfield State Museum restoration received state project no. G-1831–1867 and was approved in President Roosevelt's letter no. 1161, p. 13, receiving no. O.P.65–15–1736, letter dated November 26, 1935. Senator Matthew A. Daly to Governor Wilbur Cross, December 5, 1935; J. R. McCarl to Cross, December 16, 1935, Box 401, Folder "Whitfield House Project," WLC, CSL.

74. Evangeline Andrews to the Trustees and Members of the Special Restoration Committee of the Henry Whitfield State Museum, December 17, 1935, Box 1, Folder 6; Trustees Minutes, November 10, 1936, Box 1, Folder 3, HWHR, CSL.

75. J. Frederick Kelly to Evangeline Andrews, April 29, 1936, Box 1, Folder 14, NMI, SML.

76. Kelly, "Restoration of the Henry Whitfield House," 84–86; Index card: "Henry Whitfield House Workers—1930s Restorations," Folder "Architecture—HWH 1930–31 + 1935–37 Restoration (Kelly)," HWSM.

77. Kelly, "Restoration of the Henry Whitfield House," 79.

78. Thomas Morgan Prentice also mentioned "a curious embrasure in the southwest corner of the second floor" in "Colonial Land-Marks of New England," *Peterson Magazine* 7, no. 3 (March 1897): 259; Kelly, "Restoration of the Henry Whitfield House," 80.

79. Kelly, "Restoration of the Henry Whitfield House," 82; Smith, *The History of Guilford*, 16–17n.

80. Current research indicates that it was unlikely the Whitfields had a baffle. Author conversation with Michael McBride, HWSM curator and site administrator, May 2018; Kelly, "Restoration of the Henry Whitfield House," 87–88.

81. J. Frederick Kelly, *The Henry Whitfield House, 1639: The Journal of the Restoration of the Old Stone House, Guilford* (Hartford, CT: Prospect Press, 1939), journal entries for March 10, November 5, 17, 1936; January 26, 1937, Henry Whitfield House Restoration and Landscaping Projects, 1900–1940, Henry Whitfield House Records, RG024:001, CSL Digital Collections, https://cslib.contentdm.oclc.org.

82. Kelly, "Restoration of the Henry Whitfield House," 87–88; "Recycling in 1930," exhibit text panel at HWSM; Kelly, *The Henry Whitfield House*, journal entries for July 31, August 18, September 11, October 2, November 5, 17, 1936.

83. Andrews and Farrand spoke of ways to improve the property, which included creating a stone or wood enclosure around the house, bordering the walk leading to the street with lilacs, planting a hedge or trees to separate the street and the grounds, making a tree border around the property, and creating a "'homely' plantation of old-fashioned shrubs" around the house. Beatrix Farrand to Evangeline Andrews, November 16, 1934; Andrews to Farrand, December 4, 1934, Box 1, Folder 10, HWHR, CSL.

84. Beatrix Farrand's notes from November 29, 1935, included in Grafton M. Peberdy to Evangeline Andrews, December 3, 1935, Box 1, Folder 10, HWHR, CSL.

85. The Great New England Hurricane of September 1938 destroyed "practically all of the fine shade trees on the south side of the property," leaving only about eight trees standing. The WPA office in Hartford sent workers to Guilford from East Haven to spend two days clearing the damage. Evangeline Andrews to William Sumner Appleton, January 1, 1936, Box 1, Folder 7, HWHR, CSL; "Whitfield House, Federal Grant"; Grafton M. Peberdy to Evangeline Andrews, January 7, 1935, Henry Whitfield House Restoration and Landscaping Projects, 1900–1940, RG 024:001, CSL Digital Collections, Connecticut Digital Archive, https://col lections.ctdigitalarchive.org; Trustees Minutes, November 10, 1938, Box 1, Folder 3, HWHR, CSL.

86. Among Beatrix Farrand's most impressive works are Dumbarton Oaks in Washington, DC; Santa Barbara Botanic Garden; New York Botanical Garden; Pierpont Morgan Library; and Yale Divinity School. She also landscaped Yale University's Sterling Medical School, Marsh Botanical Garden, and the president's private garden. Diana Balmori, Diane Kostial McGuire, and Eleanor M. McPeck, *Beatrix Farrand's American Landscapes: Her Gardens and Campuses* (Sagaponack, NY: Sagapress, 1985), 8–9, 24; Judith B. Tankard, *Beatrix Farrand: Private Gardens, Public Landscapes* (New York: Monacelli Press, 2009), 163–8. Phoebe Cutler has argued that New Deal programs helped set and develop professional standards of practice for the field of landscape architecture. Cutler, *The Public Landscape of the New Deal*, 83–89.

87. Kelly, *The Henry Whitfield House*, journal entries for January 4, March 3, June 11, 1936; J. Frederick Kelly to Evangeline Andrews, April 29, 1936, Box 1, Folder 14, NMI, SML.

88. J. Frederick Kelly to Senator Matthew P. Daly, June 11, 1936; Frederick C. Norton to Daly, June 13, 1936; Norton to Governor Wilbur Cross, June 13, 1936, Box 1, Folder 7, HWHR, CSL.

89. Executive Secretary to Frederick C. Norton, June 16, 1936, Box 401, Folder "Whitfield House Project," WLC, CSL.

90. J. Frederick Kelly to Senator Matthew P. Daly, December 16, 1935, Box 401, Folder "Whitfield House Project," WLC, CSL.

91. Kelly's criticism was amended and Spencer's name removed in the published journal, however, to read, "This is properly the superintendent's duty, and it begins to look as though our wretchedly slow rate of progress were due to certain inefficiency here, as well as to the inability of the W.P.A. to furnish materials more promptly." Kelly, *The Henry Whitfield House*, journal entries for July 27, October 15, December 12, 1936; March 3, 1937.

92. The Henry Whitfield State Museum keeps a file on WPA workers; most information was provided by relatives who visited the site throughout the twentieth century. Index cards, "Henry Whitfield House Workers—1930s Restorations," Folder "Architecture—HWH 1930–31 + 1935–37 Restoration (Kelly)," HWSM.

93. J. Frederick Kelly to Evangeline Andrews, November 25, 1936, Box 1, Folder 7, HWHR, CSL.

94. Matthew A. Daly resigned on July 29, 1936, and Robert A. Hurley assumed the post on July 30. Executive Secretary to Evangeline Andrews, December 3, 1936, Box 401, Folder "Whitfield House Project," WLC, CSL.

95. Kelly, *The Henry Whitfield House,* journal entry for December 2, 1936.

96. J. Frederick Kelly to Evangeline Andrews, April 29, 1936, Box 1, Folder 14, NMI, SML.

97. Kelly, *The Henry Whitfield House,* journal entry for December 12, 1936. One of the hardworking Italian masons Kelly mentioned was probably Aldo Balestracci, a skilled worker superintendent Spencer picked up daily for six months from the nearby coastal village of Stony Creek because he "wanted a real old-fashioned stone mason" for the job. Linda Trowbridge Baxter, *The Newcomers: A Study of Immigrants to Guilford, Connecticut, 1850–1930* (Dexter, MI: Thomson-Shore, 1985), 67.

98. J. Frederick Kelly to Evangeline Andrews, July 31, 1936, Box 1, Folder 10, HWHR, CSL.

99. J. Frederick Kelly to Evangeline Andrews, June 12, 1936; Frederick Norton to Samuel L. Fisher, June 13, 1936, Box 1, Folder 7, HWHR, CSL.

100. Walter R. Steiner to Frederick C. Norton, July 5, 1936, Box 1, Folder 7, HWHR, CSL.

101. J. Frederick Kelly to Frederick C. Norton, July 7, 1936, Box 1, Folder 7, HWHR, CSL.

102. Evangeline Andrews to Walter R. Steiner, July 27, 1936, Box 1, Folder 7, HWHR, CSL.

103. Walter R. Steiner to Evangeline Andrews, August 27, 1936, Box 1, Folder 7, HWHR, CSL.

104. Everett Gleason Hill, *A Modern History of New Haven and Eastern New Haven County,* 2 vols. (New York: S. J. Clarke, 1918), 2:366.

105. Evangeline Andrews to Walter R. Steiner, July 27, 1936; J. Frederick Kelly to Andrews, August 18, 1936, Box 1, Folder 7; Kelly to Andrews, July 31, 1936, Box 1, Folder 10, HWHR, CSL.

106. For discussion of women's patriotic groups' involvement in the preservation and interpretation of domestic historic spaces, see Marling, *George Washington Slept Here,* 53–84, and West, *Domesticating History,* particularly chaps. 1–2.

107. J. Frederick Kelly to Evangeline Andrews, July 31, 1936, Box 1, Folder 10; Kelly to Andrews, August 18, 1936, Box 1, Folder 7, HWHR, CSL.

108. The state of Connecticut later transferred administration of the Henry Whitfield State Museum to the Connecticut Historical Commission in 1979, so it is possible that is when the Connecticut Society of the Colonial Dames' association with the site officially ended. Cunningham, "Whitfield, Henry, House," 12n10, 16.

109. J. Frederick Kelly to Evangeline Andrews, August 18, 27, 1936, Box 1, Folder 7, HWHR, CSL.

110. The model was built in 1934 by Henry S. Kelly, J. Frederick's brother and business partner, to display the proposed changes to the house. J. Frederick Kelly to Evangeline Andrews, August 18, 1936, Box 1, Folder 7, HWHR, CSL.

111. "Post Card Forum," *Shore Line Times*, November 4, 1937; "The Way Out of Relief," *Shore Line Times*, November 11, 1937; "Reader Expresses an Opinion," *Shore Line Times*, November 18, 1937.

112. Ruth Lee Baldwin to Evangeline Andrews, November 24, 1936, Box 1, Folder 7, HWHR, CSL.

113. For a list of gifts from private individuals, see "List of Contributions to the Whitfield House Restoration Fund," Box 1, Folder 1, HWHR, CSL, and Evangeline Andrews, "Board of Trustees of Whitfield House about Restoration," *Shore Line Times*, December 3, 1936.

114. Fairfield's Old Town Hall was built in 1720 as the town courthouse; the $24,081 WPA project constructed a new wing on either side of the building to provide office spaces and storage vaults for the town's records and "renovated" features throughout the building in "conforming colonial style." Robert A. Hurley, "Natural Beauty Spots in State Opened Up to Motorists by WPA," *Hartford Courant*, November 29, 1936; Julius F. Stone Jr. to Robert Hurley, March 10, 1937, Box 3, Folder "HS—Connecticut—Gen. Correspondence," Entry A1 764, RG 69, NACP.

115. Works Progress Administration for Connecticut, "Fairfield Meets Town Needs," in *Connecticut Work in Progress*, July 1936, 7, Internet Archive, https://archive.org/.

116. "Whitfield House," n.d., Box 3, Folder "HS—CT—Henry Whitfield Memorial Museum, Guilford," Entry A1 764, RG 69, NACP.

117. The Reorganization Commission also proposed to abolish the governing bodies of Fort Griswold and Groton Monument in Groton and the Israel Putnam Memorial Campground in Redding. For more on the reorganization plan, see Cross, *Connecticut Yankee*, 357–77.

118. Evangeline Andrews to N. D. Canterbury, April 26, 1936, Box 1, Folder 7; Trustees Minutes, November 10, 1936, Box 1, Folder 3, HWHR, CSL.

119. "Proposal to Abolish Board of Trustees of Henry Whitfield House Meets with Opposition," *Shore Line Times*, February 18, 1937.

120. Letters from organizations sent in the spring of 1937 in Box 401, Folder "Whitfield House Project," WLC, CSL.

121. Walter R. Steiner to Governor Wilbur Cross, February 16, 1937, Box 401, Folder "Whitfield House Project," WLC, CSL.

122. Roland M. Hooker to Governor Wilbur Cross, March 17, 1937; Frederic T. Murlless Jr. to Cross, February 18, 1937, Box 401, Folder "Whitfield House Project," WLC, CSL.

123. Frederick C. Norton, "Whitfield House to Remain in Care of Board of Trustees," *Shore Line Times*, April 8, 1927.

124. There was a remaining balance of $3,605.14 in unexpended federal funds. Trustees Minutes, November 10, 1938, Box 1, Folder 6, HWHR, CSL.

125. "The Whitfield House," *Shore Line Times*, May 6, 1937; "Whitfield House Open to Public," *Shore Line Times*, June 3, 1937; "Formal Opening of the Henry Whitfield House," *Shore Line Times*, July 1, 1937; "Visitors from Afar at State Museum," *Shore*

Line Times, July 8, 1937; "Many Visitors at Whitfield House," *Shore Line Times,* July 29, 1937; "Henry Whitfield State Museum," *Shore Line Times,* August 5, 1937.

126. The book, printed in 1939, included most of Kelly's journal entries word for word, but the publishing company requested to modify certain passages it felt might result in "hard feeling and possible libelous claims." The altered passages included those that painted the WPA in a poor light, such as references to superintendent Frank Spencer's "lack of efficiency" and Kelly's description of the "loafing by Yankee carpenters" on the jobsite. That same year, Kelly wrote an article about the restoration for SPNEA's quarterly journal, *Old-Time New England.* In the piece, Kelly attempted to answer some questions about the early history of the house and substantiated some long-held traditions, though admitted some "can never be answered with certainty." "Trustees of Whitfield House Name Assistant Curator at Meeting," *Shore Line Times,* May 27, 1937; Paul W. Cooley, president of Prospect Press Incorporated, to Evangeline Andrews, February 10, 1938, Box 1, Folder 8, HWHR, CSL; Kelly, "Restoration of the Henry Whitfield House," 76–77.

127. "Whitfield House Photographed," *Shore Line Times,* June 24, 1937.

128. The celebration was planned to be held on the grounds of the museum, but a storm caused the event to be moved to the First Congregational Church on the town green. Speakers included Frederick Calvin Norton, president of the Board of Trustees; Evangeline Andrews, chair of the Restoration Committee; Wilbur L. Cross, governor; Samuel H. Fisher, chair of the Tercentenary Commission; Charles M. Andrews, Pulitzer Prize–winning historian of colonial history at Yale and husband of trustee Evangeline; and William Sumner Appleton, secretary of SPNEA. "Henry Whitfield House Reconstruction to Be Commemorated Oct. 20," *Shore Line Times,* October 14, 1937; "Interesting Henry Whitfield House Program Held on Wednesday in the First Congregational Church," *Shore Line Times,* October 21, 1937.

129. "Excerpts from letter from Mr. William Sumner Appleton, October 19, 1937," in Evangeline Walker Andrews, ed., *The Henry Whitfield House: 1639: The Commemorative Exercises Held at Guilford, Connecticut, October 20, 1937, Upon the Restoration of "The Old Stone House"* (Hartford, CT: Prospect Press, 1937), 23, HWSM.

130. Evangeline Walker Andrews, "For the Reconstruction Committee," in Andrews, ed., *The Henry Whitfield House,* 13–15.

131. Arthur N. Johnson to Earl Minderman, August 21, 1940, Box 1, Folder "Connecticut," Best Project Reports, 1940" (Entry A1 755), RG 69, NACP.

132. Andrews, "For the Reconstruction Committee," 16.

133. Samuel H. Fisher, "What Connecticut Has Retained from Her Tercentenary," in Andrews, ed., *The Henry Whitfield House,* 22.

134. Federal Writers' Project of the Works Progress Administration for the State of Connecticut, *Connecticut: A Guide to Its Roads, Lore, and People* (Boston: Houghton Mifflin, 1938), 81. A photograph of the Henry Whitfield House appears between pages 88–89.

135. Federal Writers' Project of the WPA, *Connecticut*, 161.

136. Frederick C. Norton, "For the Trustees," in Andrews, ed., *The Henry Whitfield House*, 11.

137. "The Whitfield House"; "Whitfield House Open to Public"; "Formal Opening of the Henry Whitfield House"; "Visitors from Afar at State Museum"; "Many Visitors at Whitfield House"; "Henry Whitfield State Museum"; "Visitors at Henry Whitfield House," *Shore Line Times*, November 18, 1937; Minutes of the Semi-Annual Meeting of the Board of Trustees, November 11, 1937, Box 1, Folder 3, HWHR, CSL.

138. *Proceedings of Guilford Colony Tercentenary Celebration* (Guilford, CT: Shore Line Times, 1939), 103, 105–6, 974.62 G14, New Haven Free Public Library, New Haven, Connecticut.

139. David Glassberg argued that historical pageants operated "within an overarching civil-religious structure in which history furnished principles of social evolution and morality that pointed the way to future reform." Glassberg, "History and the Public," 947.

140. *Proceedings of Guilford Colony Tercentenary Celebration*, 47–48.

Chapter 4

1. "Lindbergh's Home at Little Falls Will Be Restored by Lions Club," *Saint Paul [MN] Pioneer Press*, May 26, 1935.

2. "Lindbergh Lauded for Paris Flight: Tenth Anniversary of Crossing is Marked at Celebrations Here and in Europe," *New York Times*, May 21, 1937; "Real Home Town Greets Lindbergh," *New York Times*, August 26, 1927.

3. By the early 1900s, Lindbergh Sr. and Bolander had built thirty-five houses and three commercial brick buildings in Little Falls on the west side of the Mississippi River. Bruce Larson, "Little Falls Lawyer, 1888–1906: Charles A. Lindbergh, Sr." *Minnesota History* 43, no. 5 (Spring 1973): 162, 170.

4. For a detailed architectural description of the Lindbergh House, see Cathy A. Alexander, "Charles August Lindbergh, Sr., House," National Register of Historic Places Nomination Form (Washington, DC: U.S. Department of the Interior, National Park Service, 1970), section 7, pp. 1–2, and "Lindbergh Home at Charles A. Lindbergh State Park: Pictorial Record," Binder "Lindbergh Historical Doc.," Charles Lindbergh House and Museum, Little Falls, Minnesota (hereafter CLHM).

5. Anne Morrow Lindbergh, *Locked Rooms and Open Doors: Diaries and Letters of Anne Morrow Lindbergh, 1933–1935* (New York: Harcourt Brace Jovanovich, 1974), 292.

6. Evangeline Lindbergh to Charles Lindbergh, n.d., Box 14, Folder "Lindbergh, Charles Augustus. Papers. Notes for CAL., Jr. . . . (2)," Charles August Lindbergh Papers, 1881–1967 (hereafter CALP), Charles A. Lindbergh and Family Papers, 1808–1987 (hereafter CALF), Minnesota Historical Society, Saint Paul (hereafter MHS).

7. "Lindbergh's Home Near Restoration," *New York Times*, June 28, 1931.

8. Melosh, *Engendering Culture*, 53–54.
9. Frederick Jackson Turner, *The Frontier in American History* (New York: Holt, Rinehart and Winston, 1921); Pells, *Radical Visions and American Dreams*, 24; Park and Markowitz, *Democratic Vistas*, 31.
10. Richard Slotkin, *Gunfighter Nation: The Myth of the Frontier in Twentieth-Century America* (Norman: University of Oklahoma Press, 1998), 4–5.
11. Susman, *Culture as History*, 30.
12. John William Ward, "The Meaning of Lindbergh's Flight," *American Quarterly* 10, no. 1 (Spring 1958): 9.
13. "No. 6–288: Daniel Boone's Birthplace," Box 8, "Press Releases of the NYA, 1935–1942 (with Gaps)," Folder "Release No. 6–288 Daniel Boone's Birthplace Restored by Penna. State Authorities with Aid of NYA, Oct. 31, 1938," Records of the Public Relations Section (hereafter Entry NC-35 107), Records of the National Youth Administration, Record Group 119 (hereafter RG 119), NACP. For a discussion of the NYA's restoration of the Daniel Boone homestead, see Calamia, "A New Deal for Historic Preservation," 124–34.
14. The 1940 *Boy Scouts Handbook*, printed after Lindbergh began voicing his isolationists views, had a new Norman Rockwell cover and reference to Lindbergh's "good manners" in the text was eliminated. Edward L. Rowan, *To Do My Best: James E. West and the History of the Boy Scouts of America* (Exeter, NH: Publishing Works, 2007), 155; Dixon Wecter, *The Hero in America: A Chronicle of Hero-Worship* (New York: Charles Scribner's Sons, 1941), 433. The association between Lincoln and pioneer history is explored in Erekson, *Everybody's History*.
15. Wecter, *The Hero in America*, 27–28.
16. Thomas Kessner, *The Flight of the Century: Charles Lindbergh and the Rise of American Aviation* (New York: Oxford University Press, 2010), 204.
17. Julia L. Foulkes, *To the City: Urban Photographs of the New Deal* (Philadelphia: Temple University Press, 2011), 1. Cara A. Finnegan provides an excellent discussion of mass media's presentation of Farm Security Administration photography in *Picturing Poverty*. For a collection of photographs, see Roy Emerson Stryker and Nancy Wood, *In This Proud Land: America 1935–43 as Seen in the FSA Photographs* (New York: Graphic Society, 1973).
18. Susan M. Gray, *Charles A. Lindbergh and the American Dilemma: The Conflict of Technology and Human Values* (Bowling Green, OH: Bowling Green State University Popular Press, 1985), 13; Leonard S. Reich, "From the Spirit of St. Louis to the SST: Charles Lindbergh, Technology, and Environment," *Technology and Culture* 36, no. 2 (April 1995): 365; Susman, *Culture as History*, 32–33.
19. William C. Tweed, Laura E. Soulliere, and Henry G. Law, *National Park Service Rustic Architecture: 1916–1942* (San Francisco: National Park Service, Western Regional Office, Cultural Resource Management Division, 1977), 105. Neil M. Maher draws the connection between the New Deal's conservation agenda and political support for the Roosevelt's administration's modern welfare state in *Nature's New Deal*.

20. "Crowds Pour in Home Town for Paris Hero Fete," n.d., Lindbergh Scrapbook, Box 8, CALP, CALF, MHS.

21. Bruce L. Larson, "Lindbergh's Return to Minnesota, 1927," *Minnesota History* 42, no. 4 (Winter 1970): 147–48.

22. "Lindbergh State Park, Works Progress Administration Guided Walking Tour (Summer 2010) Script," Binder "WPA and Great Depression 1930s," CLHM.

23. Larson, "Lindbergh's Return to Minnesota," 146.

24. "Lindbergh's Home at Little Falls Will Be Restored by Lions Club"; "Citizens Will Aid Lindbergh Park Project," *Little Falls [MN] Daily Transcript*, March 23, 1931.

25. "Tourists Wrecking Lindbergh's Home," *Washington Post*, July 12, 1931.

26. Sherman Leirs to Henry Breckinridge, October 11, 1928, Folder "Lindbergh State Park History: Formation and Early Restoration," CLHM.

27. Members of the Lindbergh Park Committee included local dentist Charles H. Longely, State Senator Christian Rosenmeier, family friend Martin Engstrom, automobile company owner E. A. Berg, Morrison County attorney and former mayor of Little Falls Austin Grimes, and president of the Board of Commerce A. V. Taylor. "Lindbergh's Home Near Restoration."

28. Sherman Leirs to Henry Breckinridge, October 11, 1928, Folder "Lindbergh State Park History: Formation and Early Restoration," CLHM.

29. The house had been untenanted for six years by 1931. "Lindbergh's Home Near Restoration"; Nels Nelson Bergheim to Evangeline Lindbergh, January 29, 1930 (quote), Box 13, Folder "Charles A. Lindbergh Papers (Yale Univ.) Correspondence & Misc. Papers, Jan. 1930–Nov. 1936," CALP, CALF, MHS.

30. Austin L. Grimes to Charles A. Lindbergh, August 18, 1930, Box 18, Folder "Lindbergh Papers, Correspondence, Jan.–Dec. 1931," Lindbergh Family Papers, 1808–1983 (hereafter LFP), CALF, MHS.

31. Henry Breckinridge [?] to Austin Grimes, September 25, 1930, Box 18, Folder "Lindbergh Papers, Correspondence, Jan.–Dec. 1931," LFP, CALF, MHS.

32. Anne Morrow Lindbergh became the first American woman to earn her glider's pilot license, and she often accompanied her husband on flights. Donald E. Heyhoe, "Lindbergh Four Years After," *Saturday Evening Post*, May 30, 1931.

33. Martin Engstrom to Charles Lindbergh, November 14, 1930, Box 18, Folder "Lindbergh Papers Correspondence Feb. 1929–Dec. 1930," LFP, CALF, MHS.

34. Larson, "Lindbergh's Return to Minnesota," 151.

35. Minnesota held gubernatorial elections every two years until 1962, when the governor's term increased to four years. For more on Governor Olson's political career during the New Deal, see Patterson, *The New Deal and the States*, 36, 52, 126, 129–130. For an overview of the Farmer-Labor Party as a political and economic movement in Minnesota, see Millard L. Gieske, *Minnesota Farmer-Laborism: The Third-Party Alternative* (Minneapolis: University of Minnesota Press, 1979).

36. Quoted in Barbara W. Sommer, *Hard Work and a Good Deal: The Civilian Conservation Corps in Minnesota* (Saint Paul: Minnesota Historical Society Press, 2008), 79, 99; Patterson, *The New Deal and the States*, 166.

37. Rosenmeier had previously served as Morrison County attorney and been elected to the Minnesota senate in 1922, where he became Conservative majority leader. Steven Dornfield, "Gordon Rosenmeier: The Little Giant from Little Falls," *Minnesota History* 64, no. 4 (Winter 2014): 148–51.

38. Lundeen later served the Farmer-Labor Party in the House of Representatives from 1933 to 1937, and then in the Senate from 1937 to 1940. Ernest Lundeen to Chris Rosenmeier, December 15, 1930, Box 18, Folder "Lindbergh Papers Correspondence Feb. 1929–Dec. 1930," LFP, CALF, MHS; Richard M. Valelly, *Radicalism in the States: The Minnesota Farmer-Labor Party and the American Political Economy* (Chicago: University of Chicago Press, 1989), 168–69.

39. "Lindbergh State Voted by the Senate in Minnesota," *New York Times*, February 6, 1931; Martin Engstrom to Henry Breckinridge, March 5, 1931, Box 18, Folder "Lindbergh Papers, Correspondence, Jan.–Dec. 1931," LFP, CALF, MHS; "Lindbergh State Park Bill Passed by House," *Little Falls Daily Transcript*, March 7, 1931; Roy W. Meyer, *Everyone's Country Estate: A History of Minnesota's State Parks* (Saint Paul: Minnesota Historical Society Press, 1991), 93.

40. "Citizens Will Aid Lindbergh Park Project."

41. Charles Lindbergh to Martin Engstrom, February 23, 1932, Box 18, Folder "Lindbergh Papers, Correspondence, 1932–1979," LFP, CALF, MHS.

42. "Deliver Lindbergh Deed to Minnesota," *Little Falls Daily Transcript*, April 1, 1931.

43. Stafford King served as state auditor until 1969. Rolf T. Anderson, "Minnesota State Park CCC/WPA/Rustic Style Historic Resources," National Register of Historic Places Multiple Property Listing (Washington, DC: U.S. Department of the Interior, National Park Service, 1989), section E, pp. 1–4.

44. "Engstrom Named Lindbergh Park Head," *Little Falls Daily Transcript*, April 28, 1931; "Sutliff is Named Park Caretaker," *Little Falls Daily Transcript*, May 1, 1931; "Lindbergh Home Being Repaired," *Leader-Telegram* (Eau Claire, WI), May 21, 1936.

45. "King Here on Park Details," *Little Falls Daily Transcript*," April 27, 1931.

46. "Work on the Lindbergh Home in Full Swing," *Little Falls Daily Transcript*, July 11, 1931.

47. "Lindbergh's Home Near Restoration"; "Work on the Lindbergh Home in Full Swing"; "Lindbergh Home Assuming Former Neat Appearance; Will Be Opened September 1," *Little Falls Daily Transcript*, August 13, 1931.

48. David Welky, *Everything Was Better in America: Print Culture in the Great Depression* (Urbana: University of Illinois Press, 2008), 29.

49. Joyce Milton, *Loss of Eden: A Biography of Charles and Anne Morrow Lindbergh* (New York: HarperCollins, 1993), 237, 299.

50. Wecter, *The Hero in America*, 27–28.

51. "Flight Anniversary Marred by Tragedy; Lindbergh Park Dedication Postponed," *Little Falls Daily Transcript*, May 20, 1932.

52. The Lindberghs returned to the United States in 1939 when Lindbergh began publicly campaigning for the isolationist America First Committee.

53. "Lindbergh's Home at Little Falls Will Be Restored by Lions Club."

54. "Lindberghs Circle over Home Town in Fast Plane as Surprise Visit Ends," *Little Falls Daily Transcript*, July 29, 1935; "City Welcomes Lindberghs in Surprise Visit," *Little Falls Daily Transcript*, July 27, 1935; "Lindbergh Off in Plane after Trip by Auto," *Little Falls Daily Transcript*, August 31, 1935; "Lindbergh Revisits Minnesota Friends," *New York Times*, August 31, 1935.

55. In 1971, the Department of Conservation's name was changed to the Department of Natural Resources. Sommer, *Hard Work and a Good Deal*, 99–100.

56. Maher, *Nature's New Deal*, 13, 70–71.

57. Anderson, "Minnesota State Park CCC/WPA/Rustic Style Historic Resources," section E, p. 24.

58. One WPA worker described Christgau as "kind of a right-wing bureaucrat," while Mabel S. Ulrich, head of the Minnesota FWP, appraised him as "a young liberal" who directed the expenditure of relief funds meticulously and judiciously. Interview with Elizabeth Hoff Bruce, June 26, 1989, Twentieth Century Radicalism in Minnesota Oral History Project, OH 30, AV1990.228.7, MHS Collections Online, https://www.mnhs.org; Mabel S. Ulrich, "Salvaging Culture for the WPA," *Harpers Magazine*, May 1939, 662.

59. "Like Son, Like Father," *Salt Lake [UT] Telegram*, October 11, 1935.

60. "Project to Make County Work Center," *Little Falls Daily Transcript*, September 11, 1935.

61. Over twenty state parks in Minnesota received WPA funds. To see a full list of state parks that received federal assistance from the CCC, WPA, and NYA, see Anderson, "Minnesota State Park CCC/WPA/Rustic Style Historic Resources," section E, p. 17.

62. The Charles A. Lindbergh State Park received $64,171 from the WPA and $7,132.55 from the state of Minnesota. Anderson, "Minnesota State Park CCC/WPA/Rustic Style Historic Resources," section E, pp. 27–28.

63. Harold W. Lathrop to Victor Christgau, March 24, 1937, Box 6, Folder "HS—MN—Lindbergh Home & Park," Entry A1 764, RG 69, NACP; "Lindbergh Park Program Approved on Anniversary," *Little Falls [MN] Weekly Transcript*, May 21, 1936; "Lindbergh Park Project Approved," *Little Falls [MN] Herald*, May 27, 1936.

64. Grace Nute earned her doctorate in history in 1921 from Harvard University, where she worked with preeminent historian of the frontier thesis Frederick Jackson Turner.

65. Nute spent years researching the Lindberghs, which included a trip to Sweden that was financed in part by Charles Lindbergh. MHS mysteriously ended its support of the project, and the book on the Lindbergh family was never published. "Charles August Lindbergh and Family" Manuscript Collections Finding Aid, Gale Library, MHS, https://www.mnhs.org.

66. Throughout his life, Lindbergh remained interested in the park named for his father and continued to provide observations of the home in which he grew up. In the 1950s, Lindbergh wrote to Arch Grahn of the MHS with surprisingly detailed

remembrances of his boyhood home, giving additional descriptions of the wall color of various rooms and furniture pieces. He also offered to send family items from his personal storage to the MHS. Charles Lindbergh to Arch Grahn, June 13, 1950, Binder "Lindbergh Historical Doc.," CLHM. See also "Comments and Observations on the Review of the Lindbergh Property with Charles Lindbergh, Jr. in Charles Lindbergh State Park—April 19 and 20," April 27, 1966, Folder "Records Concerning the Charles A. Lindbergh House, 1955–1970," Department of Natural Resources, Parks and Recreation, Records of the Lindbergh House, 103.E.19.10F, MHS.

67. Charles Lindbergh to Grace Nute, July 27, 1937; March 23, 1938, Box 2, Folder "Lindbergh Papers, Grace Lee Nute Correspondence—Charles A. Lindbergh Jan.–Dec. 1937," LFP, CALF, MHS.

68. Charles Lindbergh to Grace Nute, January 12, 1938, Box 2, Folder "Lindbergh Papers. Grace Lee Nute Correspondence: Charles A. Lindbergh, Jan.–Dec. 1938," LFP, CALF, MHS. Lindbergh later published his recollections in *Boyhood on the Mississippi: A Reminiscent Letter* (Saint Paul: Minnesota Historical Society, 1972).

69. The Minnesota Historical Society staff eventually removed the front steps to give the house a more authentic appearance. Charles Lindbergh to Theodore Blegen, September 21, 1936, Box 2, Folder "Lindbergh Papers, Grace Lee Nute Correspondence: Lindbergh Home, 1936," LFP, CALF, MHS; "WPA Exhibit Text," March 28, 2008, Binder "WPA & Great Depression 1930s," CHLM.

70. Grace Nute to Charles Lindbergh, December 27, 1938; Lindbergh to Nute, February 2, April 3, 1939, Box 2, Folder "Lindbergh Papers, Grace Lee Nute Correspondence: Charles A. Lindbergh Feb.–Dec. 1939," LFP, CALF, MHS.

71. Charles Lindbergh to Grace Nute, June 6, August 8, 1939, Box 2, Folder "Lindbergh Papers, Grace Lee Nute Correspondence: Charles A. Lindbergh Feb.–Dec. 1939," LFP, CALF, MHS.

72. Evangeline Lindbergh to Charles Lindbergh, n.d., Box 14, Folder "Lindbergh, Charles Augustus. Papers. Notes for CAL., Jr. . . . (2)," CALP, CALF, MHS.

73. Additional pieces of Evangeline Land Lindbergh's furniture were moved to the site after her death in 1954. Charles Lindbergh to Theodore Blegen, June 12, 1936, Box 2, Folder "Lindbergh Papers, Grace Lee Nute Correspondence: Lindbergh Home, 1936," LFP, CALF, MHS.

74. "Plan Mural at Lindbergh Home in Park," *Little Falls Daily Transcript*, June 11, 1936; "WPA Sets $800 Aside to Furnish Home of Lindbergh," *Little Falls Daily Transcript*, January 22, 1936.

75. Charles Lindbergh to Theodore Blegen, June 12, 1936; Blegen to Lindbergh, October 17, 1936, Box 2, Folder "Lindbergh Papers, Grace Lee Nute Correspondence: Lindbergh Home, 1936," LFP, CALF, MHS.

76. Charles Lindbergh expressed to Grace Nute that two things he was "most anxious to have done" were the planting of small pine trees and reforesting of the valley on the side of the house facing the river, whose trees had been eradicated by erosion. Both wishes were carried out by WPA workers. Charles Lindbergh to

Grace Lee Nute, October 29, 1936, Box 2, Folder "Lindbergh Papers, Grace Lee Nute Correspondence: Lindbergh Home, 1936," LFP, CALF, MHS.

77. Tweed, Soulliere, and Law, *National Park Service Rustic Architecture*. For more on Downing and Olmsted, see William H. Tishler, ed., *American Landscape Architecture: Designers and Places* (Washington, DC: National Trust for Historic Preservation, 1989), 31–33, 38–41, and William C. Tweed, *Parkitecture: A History of Rustic Building Design in the National Park System: 1916–1942* (San Francisco: National Park Service, Western Regional Office, Cultural Resource Management Division, 1978), 129.

78. Merrill Ann Wilson, "Rustic Architecture: The National Park Style," *Trends* (July–September 1976): 4–5.

79. Constance Rourke, "American Art: A Possible Future," *Magazine of Art*, 28, no. 7 (July 1935): 390–405; Musher, *Democratic Art*, 27–28; McLerran, *A New Deal for Native Art*.

80. Albert H. Good, *Park Structures and Facilities* (Washington, DC: National Park Service, Department of the Interior, 1935), 3–4. The first edition sold out and an expanded three-volume edition titled *Park and Recreation Structures, Parts I, II, and III, Administration and Basic Service Facilities* was published three years later in 1938.

81. Regional adaptions of Rustic style appeared in other parts of the country as well; in the Southwest, workers constructed pueblos (adobe houses) and Spanish Colonial structures in NPS parks. Anderson, "Minnesota State Park CCC/WPA/Rustic Style Historic Resources," section E, pp. 34–36.

82. Limestone was most abundant in the southern portion of the state; dark basalt rock and sandstone in the east; quartzites in the southwest; granite along Lake Superior shorelines; and fieldstone in the west, north, and southwest. Anderson, "Minnesota State Park CCC/WPA/Rustic Style Historic Resources," section E, p. 40.

83. For example, the T-shaped log kitchen shelter at the Lindbergh State Park was nearly identical to the shelter at Bemidji State Park, and the shelter pavilion at Lake Bronson State Park mirrored the one at Sibley State Park. Sommer, *Hard Work and a Good Deal*, 100.

84. Other excellent examples of Rustic-style architecture in Minnesota's state parks include the River Inn at Jay Cooke State Park, the Forest Inn at Itasca State Park, and the shelter building and bathhouse at Whitewater State Park. Anderson, "Minnesota State Park CCC/WPA/Rustic Style Historic Resources," section E, pp. 38–41; Sommer, *Hard Work and a Good Deal*, 103, 105.

85. Theodore F. Meltzer, "Lindbergh State Park," *Minnesota Conservationist*, November 1937, 16.

86. The interior of the park shelter featured a large stone fireplace, sink, and four cast-iron wood stoves. "Improvements Being Made by the Works Progress Administration in Chas. A. Lindbergh State Park," March 27, 1937," Box 6, Folder "HS—MN—Pictorial Report of Improvements Made by WPA—Lindbergh Park," Entry A1 764, RG 69, NACP.

87. For more information on the Lindbergh park's WPA structures, see "Charles A. Lindbergh State Park Management Plan—June 1998," CLHM.

88. "WPA Exhibit Text," March 28, 2008, Binder "WPA & Great Depression 1930s," CHLM.

89. "$16,000 Fund to Complete Work at Park," *Little Falls Daily Transcript,* January 26, 1938; "Improvements for Lindbergh Park," *Little Falls Herald,* January 28, 1938; "Allot $5,359 on State Park," *Little Falls Daily Transcript,* March 10, 1941.

90. Norman Christensen, "Floods in Minnesota Give New Emphasis to Old Warning: Stop Erosion or Lose Farm Lands," *Minneapolis Tribune,* May 17, 1936.

91. "To Wage War on Tree Diseases," *Little Falls Herald,* April 17, 1936.

92. "Improvements Being Made by the Works Progress Administration in Chas. A. Lindbergh State Park," March 27, 1937.

93. Sarah T. Phillips argues that conservation policy was at the center of New Deal recovery efforts and explores agricultural reform across different New Deal programs in rural America, including the Tennessee Valley Authority and the Agricultural Adjustment Agency, in *The Land, This Nation: Conservation, Rural America, and the New Deal* (New York: Cambridge University Press, 2007).

94. For reactions to the CCC from enrollees and the public, see Sommer, *Hard Work and a Good Deal.*

95. This work included the planting of more than 21,000 trees of various species, including white, jack, and Norway pine; white spruce; red, northern pine, and burr oak; black and white ash; hackberry, white, red, and rock elm; basswood; butternut; birch; hard and soft maple; thornapple; and wild plum. Meltzer, "Lindbergh State Park," 26.

96. Other WPA projects in Little Falls included the addition of water mains throughout town, road resurfacing, tree planting, and the establishment of the Morrison County Historical Society, based in Little Falls, which initiated a project to record biographies of county residents. Mary Warner, "The W.P.A.—Conversations with Jan [Warner]," Morrison County Historical Society, http://morrisoncountyhistory .org; "Dream of Good Water in City Seen Reality," *Little Falls Daily Transcript,* April 14, 1936; "WPA to Organize Historical in Morrison County," *Little Falls Weekly Transcript,* July 25, 1936; "Kasparek Heads Historical Group," *Little Falls Herald,* September 4, 1936.

97. Clarence Tuller, who trained as a teacher, stopped working for the WPA in 1938 when he was hired by the Little Falls schoolboard. "Clarence Tuller Interview," March 20, 2007, Binder "WPA & Great Depression 1930s," CLHM.

98. David Benson, *Stories in Log and Stone: The Legacy of the New Deal in Minnesota State Parks* (Saint Paul: Minnesota Department of Natural Resources, Division of Parks and Recreation, 2002), 60.

99. Victor Christgau to David K. Niles, February 26, 1937, Box 6, Folder "HS—MN— Lindbergh Home & Park," Entry A1 764, RG 69, NACP.

100. At Fort Ridgely, CCC enrollees constructed new buildings of quarried Morton rainbow granite, restored the original commissary building, and conducted

archaeological excavation under the direction of the MHS and on the recommendation of the regional historian of the NPS. The CCC also reconstructed the Fond du Lac Trading Post at the Jay Cooke State Park and the North West Company Fur Post at Grand Portage State Park. Sommer, *Hard Work and a Good Deal*, 106, 119–21. The Faribault House restoration was sponsored by the Daughters of the American Revolution and the State Highway Department. It began as a FERA project but later received funding from the WPA and the Public Works Administration. Ann Marcaccini and George Woytanowitz, "House Work: The DAR at the Sibley House," *Minnesota History* 55, no. 5 (Spring 1997): 186–201.

101. Anderson, "Minnesota State Park CCC/WPA/Rustic Style Historic Resources," section E, p. 24. The WPA restored the Longfellow House, a two-thirds-scale replica of the poet's house in Cambridge, Massachusetts, in 1937, and rehabilitated it into the Longfellow Community Library. Albert D. Wittman, *Architecture of Minneapolis Parks* (Charleston, SC: Arcadia, 2010), 20.

102. Victor Christgau to David K. Niles, November 18, 1937, Box 6, Folder "HS—MN—Lindbergh Home & Park," Entry A1 764, RG 69, NACP. Ralph D. Brown sent examples of other restoration projects in Minnesota to be considered in the program to Henry Alsberg, national director of the Federal Writers' Project. Brown's list included Fort Ridgely near Fairfax, the Faribault House in Mendota, and the J. Harley Smith house and park in Montevideo. Ralph D. Brown to Henry G. Alsberg, March 16, 1937, Box 6, Folder "HS—MN—Gen Correspondence," Entry A1 764, RG 69, NACP; Sommer, *Hard Work and a Good Deal*, 119–21.

103. Christgau erroneously called the property "Lindbergh's birthplace." Julius F. Stone Jr. to Victor Christgau, March 4, 1937; Christgau to Stone, March 30, 1937, Box 6, Folder "HS—MN—Lindbergh Home & Park," Entry A1 764, RG 69, NACP.

104. Harold Lathrop cited in Meltzer, "Lindbergh State Park," 16.

105. Federal Writers' Project of the Works Progress Administration, editorial notes for "Minnesota Recreation Guide," n.d., Box 226, Folder "WPA Writers' Project 'Minnesota Recreation Guide,'" Minnesota Works Progress Administration Collection, c. 1935–43 (hereafter MWPAC), MHS.

106. Amy Lonetree, *Decolonizing Museums: Representing Native America in National and Tribal Museums* (Chapel Hill: University of North Carolina Press, 2012), 35. For discussion of the efforts of the CCC–Indian Division, see McLerran, *A New Deal for Native Art*, 200–223.

107. Ulrich, "Salvaging Culture for the WPA," 657.

108. Mabel S. Ulrich to Henry Alsberg, March 8, 1937, cited in Kenneth E. Hendrickson Jr., "The WPA Federal Art Projects in Minnesota, 1935–1943," *Minnesota History* 53, no. 5 (Spring 1993): 176–77. A focus on the mythologized heroic frontier past was prevalent in other federal cultural programs too. For example, the Federal Theatre Project and the Treasury Section of Fine Arts presented Abraham Lincoln as the "exemplary frontiersman." Melosh, *Engendering Culture*, 35. Keith A. Erekson examines the activities of the Southwestern Indiana Historical Society in researching and popularizing Lincoln's memory in the 1920s and 1930s in *Everybody's History*.

109. Federal Writers' Project of the Works Progress Administration, *Minnesota: A State Guide* (New York: Viking Press, 1938), 3–4.

110. Karal Ann Marling, *Blue Ribbon: A Social and Pictorial History of the Minnesota State Fair* (Saint Paul: Minnesota Historical Society Press, 1990), 210.

111. Curtis Erickson, "The Father of the Lone Eagle," Box 232, Folder "WPA Writers' Project, Curtis Erickson, 'The Father of the Lone Eagle,'" MWPAC, MHS.

112. Charles A. Lindbergh, *The Autobiography of Values* (New York: Harcourt Brace Jovanovich, 1978), 4.

113. "Reeve Lindbergh, Introductory Remarks at Charles A. Lindbergh Memorial Lecture," St. Cloud State University, June 17, 1981, Box 1, Folder "Charles A. Lindbergh Papers, Family Histories, Reeve Lindbergh Speech at St. Cloud State Univ. June 17, 1981," LFP, CALF, MHS.

114. Lindbergh, *Locked Rooms and Open Doors*, 290, 292–93.

115. Anne Morrow Lindbergh's remarks at the special program observing the fiftieth anniversary of Charles A. Lindbergh's flight, May 22, 1977, cited in John William Ward, "Lindbergh and the Meaning of American Society," *Minnesota History* 45, no. 7 (Fall 1977): 291.

116. Ward Morehouse, "Lindy's Name Has Cash Value to Little Falls, His Home Town, but It Isn't Commercialized," *Minneapolis Tribune*, May 20, 1936; "A Columnist Looks at Little Falls," *Little Falls Weekly Transcript*, May 20, 1936.

117. Charles Lindbergh to Grace Nute, February 2, 1939, Box 2, Folder "Lindbergh Papers, Grace Lee Nute Correspondence: Charles A. Lindbergh Feb.–Dec. 1939," LFP, CALF, MHS.

118. Charles Lindbergh to Martin Engstrom, October 30, 1936; Engstrom to Lindbergh, July 11, 1937, Binder "WPA & Great Depression 1930s," CLHM.

119. Anton Janson, "Public Forum: Park Breezes," *Little Falls Herald*, June 16, 1939.

120. "Address at the Dedication of Lindbergh State Park," n.d., Box 3, Folder "Charles A. Lindbergh Papers. Address of the Dedication of Lindbergh State Park, Undated," LFP, CALF, MHS.

121. Valelly, *Radicalism in the States*, 101–2; Elmer A. Benson, "Politics in My Lifetime," *Minnesota History* 74, no. 4 (Winter 1980): 154. Benson called WPA state administrator Victor Christgau "a thorn in my side while I was governor . . . a right-wing, former Republican congressman who . . . had ambitions to switch to the Democrats and run for governor against me." For a New Dealer's account of Christgau's dismissal from the WPA, see Ulrich, "Salvaging Culture for the WPA," 662–63. Letters in support for and against Christgau sent to Harry Hopkins of the WPA, along with letters between Governor Benson and Christgau, can be found in Box 1611, Folder 610, Minnesota July–Dec. 1936, Administrative and Operational Correspondence Relating to Minnesota, 1935–44 (Entry PC-37 12-25), RG 69, NACP.

122. Roger Butterfield, "Lindbergh: A Stubborn Young Man of Strange Ideas Becomes Leader of Wartime Opposition," *Life*, August 11, 1941, 65–75.

Chapter 5

1. "Bexar_La Villita Under the Stars_captions," Bexar County, Texas, Box 78, RG 69-NS, WPA Division of Information State Files, 1934–42, Texas and the WPA (Texas State Parks) Flickr, https://www.flickr.com.

2. "First of Its Kind, City Council Passes 'Picture Book' Ordinance Providing 'La Villita,'" *San Antonio [TX] Light*, October 12, 1939.

3. Sources often list 1722 as the year La Villita was founded, and that is the date used throughout the NYA project, but its accuracy is under debate. "The 1722 Question," March 1, 2022, Folder "La Villita Gen'l 1817–1940," San Antonio Conservation Society, San Antonio, Texas (hereafter SACS).

4. Laura R. Barraclough, *Charros: How Mexican Cowboys Are Remapping Race and American Identity* (Oakland: University of California Press, 2019), 215n5.

5. Mae Ngai, *Impossible Subjects: Illegal Aliens and the Making of Modern America* (Princeton, NJ: Princeton University Press, 2004), 7.

6. Sean P. Cunningham, *Bootstrap Liberalism: Texas Political Culture in the Age of FDR* (Lawrence: University Press of Kansas, 2022), 85.

7. Judith Kaaz Doyle wrote that Maverick was "a staunch champion of civil rights in the abstract" but had to contend with the legacy of racial prejudices and traditions as a southern liberal. For example, he limited Black political participation in San Antonio by supporting the white primary. For more on Maverick's liberal image and his political record surrounding race, see Doyle, "Maury Maverick and the Racial Politics in San Antonio, Texas, 1838–1941," *Journal of Southern History* 53, no. 2 (May 1987): 194–95, 202 (quote), and Stuart L. Weiss, "Maury Maverick and the Liberal Bloc," *Journal of America History* 57 (March 1971): 881–95.

8. Richard B. Henderson, *Maury Maverick: A Political Biography* (Austin: University of Texas Press, 1970), 199.

9. For more on the Democratic Texas coalition in Congress, see George M. Cooper, "South Texans in Washington during the New Deal," in Jordan and Cooper, eds., *Conflict and Cooperation*, 28–47.

10. The Democratic machine in San Antonio was controlled by then-Mayor Charles Kennon (C. K.) Quin, Owen Kilday, and Valmo Bellinger, who campaigned to replace Maverick with Owen's brother Paul Kilday. Kilday went on to represent San Antonio's 20th Congressional District until 1961. Doyle, "Maury Maverick and the Racial Politics in San Antonio," 203; Cunningham, *Bootstrap Liberalism*, 157–59.

11. Henderson, *Maury Maverick*, 191.

12. In the 1930 census only, Mexicans' racial status had been changed from "white" to "other races." Laura Hernández-Ehrisman, *Inventing the Fiesta City: Heritage and Carnival in San Antonio* (Albuquerque: University of New Mexico Press, 2008), 79–81; Zaragosa Vargas, *Labor Rights Are Civil Rights: Mexican American Workers in Twentieth-Century America* (Princeton, NJ: Princeton University Press, 2005), 129, 146.

13. Maury Maverick, untitled article for *Look Magazine*, October 3, 1939, 18, Box 2L70, Folder "Literary Production, 1930–1939," Maury Maverick, Sr., Collection, 1769–1954, 1989 (hereafter MMC), Dolph Briscoe Center of American History, University of Texas at Austin (hereafter DBCAH).

14. This was a draft of a booklet presented to the Rockefeller Foundation. "La Villita," City of San Antonio, Texas, National Youth Administration of Texas, September 1939, Folder "La Villita—Ordinance," SACS.

15. Maury Maverick, untitled story for the *Christian Science Monitor*, March 24, 1941, 5, Box 2L70, Folder "Literary Production, 1940–1949," MMC, DBCAH.

16. "Ernie Pyle Says; a Block of San Antonio Slums Is Converted into a Show Place and a Home for Handicraft," *Ogdensburg [NY] Advance-News*, December 21, 1939. Maverick's support for La Villita as a bridge between the Americas joined an already strong Pan-American movement in San Antonio organized by the local chapter of the Pan American Round Table, a women's activist organization formed in 1916. See Dina Berger, "Raising Pan Americans: Early Women Activists of Hemispheric Cooperation, 1916–1944," *Journal of Women's History* 27, no. 1 (2015): 40–45.

17. John H. Sprinkle Jr., "'History Is as History Was, and Cannot Be Changed': Origins of the National Register Criteria Consideration for Religious Properties," *Buildings and Landscapes* 16, no. 2 (Fall 2009): 3; Fisher, *Saving San Antonio*, 164; Henderson, *Maury Maverick*, 94.

18. Lewis F. Fisher, "The Preservation of San Antonio's Built Environment," in *On the Border: An Environmental History of San Antonio*, ed. Char Miller (Pittsburgh: University of Pittsburgh Press, 2001), 159.

19. Maury Maverick, "The Story of Villita," n.d., San Antonio Archival File, Villita, La, Brochure, "Villita" flyer, San Antonio Public Library, San Antonio, Texas (hereafter SAPL).

20. Local newspaper articles frequently used the term "pet project" to refer to the mayor's attention to La Villita. "Interview with Terrell Dobbs Maverick Webb, 1977," 28, Institute of Texan Cultures Oral History Collection, OHT 721.9 W368, University of Texas San Antonio (UTSA) Special Collections Digital Collections, https://digital.utsa.edu.

21. The SACS bought the Villita Art Gallery, originally built in 1850 for Jeremiah Dashiell, in 1942 for $7,000 when the city would not purchase it for the La Villita project. Hernández-Ehrisman, *Inventing the Fiesta City*, 95–96; Fisher, *Saving San Antonio*, 198, 221.

22. For an overview of the NYA in Texas, see Scogin-Brincefield, "'The Yield on This Investment Should Be High,'" 65–77.

23. Maverick's widow, Terrell Webb, called Aubrey Williams "one of our most intimate friends. . . . Not only were they [Maury Maverick and Aubrey Williams] political friends, we were very close personally." "Interview with Terrell Dobbs Maverick Webb," 3–4.

24. "Maverick as Mayor 'Gets Senate Experience,'" *Washington Evening Star*, July 23, 1939; "S.A. Village Funds Are Granted," *San Antonio Light*, July 24, 1939.

25. "Federal Security Agency, National Youth Administration, Division of Information, District 2155, PR-33," October 2, 1939, Box 1, Folder 3: "Press Releases of the NYA, 1935–1942," Entry NC-35 107, RG 119, NACP.

26. Having purchased the property in 1930 to place new generator units for a station plant, the company later decided to locate them elsewhere and tenants continued to occupy their La Villita homes, mostly rent-free. The company did not keep up with maintenance or taxes, and despite being downtown, appraised property values were low as the area continued to fall into disrepair during the company's tenure. Linda Claire Scarbrough, "La Villita: Convergence of Life, Politics and Art," May 3, 1996, 17, School of Architecture and Planning, University of Texas at Austin, Call No. 917.64351, SACS.

27. "Colorful Spanish Village to Be Restored on Villita Street Site," *San Antonio Light*, July 18, 1939. In addition to the equal cash payments of $55,097.73, the city ceded to the company two dead-end streets, one in the street railway "car barn" area east of San Pedro Avenue across from San Pedro Park, and the other in the gas plant area in the 500 block of South Salado Street. A full account of how the city of San Antonio acquired property for La Villita in the 1930s can be found in Wilbur L. Matthews, "How the City Acquired La Villita," *Our Heritage* 34, no. 2 (Winter 1992–93): 10–12.

28. Emma Tenayuca had decided not to hold the meeting of the Communist Party, but her husband and fellow communist Homer Brooks went ahead with it anyway. One resident remembered the protests having to be settled down with fire hoses. In the evening, rioters hanged Mayor Maverick in effigy from a lamppost at City Hall. Cunningham, *Bootstrap Liberalism*, 173; Don Politico, "During 1939 It Was a Tumultuous City Hall," *San Antonio Light*, January 1, 1940; Victor Jaeggli, oral history interview 1 (I), September 4, 1979, by Michael L. Gillette, p. 13, LBJ Library Oral Histories, LBJ Presidential Library, https://www.discoverlbj.org; Vargas, *Labor Rights Are Civil Rights*, 146.

29. The timing of the returning indictments prevented Maverick from appearing in an NYA film about the restoration of La Villita. "Poll Tax Probe Just Started, Prosecutor Says, Following Indictment of Maverick," *San Antonio [TX] Express*, October 17, 1939.

30. The city also copyrighted the song "In Old Villita" and later sought copyright for the word "Villita." At the same time, the city clerk's office established the La Villita Register to house all documents concerning the project and changed the name of Womble Alley to King Philip V Street after the Spanish king. "City to Re-Create La Villita at Council Meeting Today," *San Antonio Express*, October 12, 1939; "City Law to Be Colorful," *San Antonio Light*, October 20, 1939; "City Gets Copyright to La Villita Song," *San Antonio Express*, January 16, 1940; "City Protects Its Villita with Copyright," *San Antonio Express*, February 4, 1940.

31. "First of Its Kind, City Council Passes 'Picture Book' Ordinance," *San Antonio Light*, October 12, 1939; "City Law to Be Colorful."

32. "The Villita Ordinance," October 12, 1939, Box 6, Folder "Texas Information Reports," Correspondence with State Officials Concerning Publicity and Other Informational Matters, 1938–1942 (hereafter Entry NC-35 86), RG 119, NACP.

33. Maverick also had the WPA Federal Writers' Project translate inscriptions for the Cos House and Caxias House into Spanish and Portuguese. Documents from January–February 1941, Box 4J200, Folder 4: "WPA Bexar Co.—San Antonio—La Villita," Works Progress Administration Records, 1935–45 (hereafter WPAR), DBCAH.

34. "Open House at City Hall," *San Antonio Light*, December 26, 1939.

35. In late October 1939, the city bought land described as block 900, lot E (420 Villita Street) from Finton C. Rice and Kathryn R. Wright for $2,200 in cash, with "the assumption of a $2,500 note against the property and the transfer of another lot owned by the city." The city then paid $1,800 for the transfer of property from Louis Henry Theis in February 1940. "Movie to Aid Financing of Villita Project," *San Antonio Express*, October 29, 1939; "La Villita Gets Last of Land," *San Antonio Express*, February 6, 1940; "The Villita Ordinance."

36. "Story by Maury Maverick for the *Christian Science Monitor*," March 24, 1941, 4, Box 2L70, Folder "Literary Production, 1940–1949," MMC, DBCAH; "Interview with Louis Lipscomb, 1976," 1–2, 9–10, Institute of Texan Cultures Oral History Collection, OHT 320.9764351 L767, UTSA Special Collections Digital Collections.

37. "San Antonio's Maury Maverick to See for Himself at Ranger," *Texas NYA Digest*, December 1939, Box 132, NYA Files of Processed and Printed Material, 1931–42, NYA Publications Files (hereafter Entry NC-35 330), RG 119, NACP; "Historical Spot in San Antonio Is Being Restored," *Corsicana [TX] Daily*, November 25, 1939.

38. Pyle merely repeated the description of squalor and human misery from the NYA Progress Report published the previous month, as did architect Arch B. Swank in his 1940 essay on La Villita when describing it as "a blighted area of unbelievable squalor." Ernie Pyle, "San Antonio Slum District Becomes One Block of Hope," *El Paso [TX] Herald Post*, December 22, 1939; A. B. Swank, "The Villita Project," *Southwest Review* 25, no. 4 (July 1940): 397.

39. Aubrey Neasham, "Special Report on 'La Villita' (In San Antonio, Texas)," February 1940, Folder "La Villita History Gen'l 1817–1940," SACS.

40. Texas Writers' Project of the Works Progress Administration, *Old Villita* (San Antonio: Clegg, 1939), Box 6, Folder "Texas Information Reports," Entry NC-35 86, RG 119, NACP.

41. Maury Maverick, untitled article for *Look Magazine*, October 3, 1939, 18, Box 2L70, Folder "Literary Production, 1930–1939," MMC, DBCAH.

42. Vargas, *Labor Rights Are Civil Rights*, 39.

43. For example, agricultural workers were not covered by the National Labor Relations Act of 1935, the Social Security Act of 1935, or the Fair Labor Standards Act of 1938. Ngai, *Impossible Subjects*, 132, 136.

44. Mexican Americans were not legally allowed to buy houses or land outside of the West Side, unless they claimed to be Spanish instead of Mexican. Robert B. Fairbanks, "Public Housing for the City as a Whole: The Texas Experience, 1934–1955," *Southwestern Historical Quarterly* 103, no. 4 (April 2000): 409; Donald L. Zelman, "Alazan-Apache Courts: A New Deal Response to Mexican Housing Conditions in San Antonio," *Southwestern Historical Quarterly* 87, no. 2 (October 1983): 124–26; Hernández-Ehrisman, *Inventing the Fiesta City*, 79.

45. Fairbanks, "Public Housing for the City as a Whole," 409.

46. Zelman, "Alazan-Apache Courts," 126.

47. Barraclough, *Charros*, 70.

48. For an illuminating statistic, in 1934 the Mexican community represented less than 40 percent of the city's population but accounted for 69 percent of tuberculosis-related deaths, and over 70 percent in 1937. Zelman, "Alazan-Apache Courts," 129.

49. "Plan New Shed for City Market," *San Antonio Light*, December 17, 1939.

50. "S.A. Death Rate Figures High," *San Antonio Light*, October 12, 1939.

51. It is important to note that San Antonio officials *chose* to separate New Deal housing projects by race—the U.S. Housing Authority did not mandate it—thereby contributing to a national trend. Victoria Courts was built for whites and Lincoln Heights and Wheatley Courts for African Americans. By the end of World War II, more than 2,554 family units of public housing had been built in San Antonio; of those units, Mexican Americans received 1,180, Anglos 796, and African Americans 578. Fairbanks, "Public Housing for the City as a Whole," 410, 413; Zelman, "Alazan-Apache Courts," 132, 138, 141; Vargas, *Labor Rights Are Civil Rights*, 187.

52. "La Villita Progress Report," November 1939, Box 6, Folder "Texas Information Reports," Entry NC-35 86, RG 119, NACP.

53. "La Villita Progress Report."

54. "Interview with Louis Lipscomb," 10. A WPA narrative titled "Villita" provides the following description: "Families were moved from their dreary surroundings, many of them eventually to find new hope and better life in the FHA apartments built at Alazan and Apache Courts." WPA, "Villita," 7, Box 4J200, Folder 4: "WPA Bexar Co.—San Antonio—La Villita," WPAR, DBCAH.

55. Architect O'Neil Ford later admitted to it being a mistake to demolish the two commercial brick buildings dating to c. 1890. At the time, he said, "we thought we needed the space for open place and trees." "City to Begin Work Next Week on $100,000 Spanish Village," *San Antonio Express*, September 9, 1939; "Spanish Village Draft Presented," *San Antonio Light*, September 17, 1939; "Interview with O'Neil Ford, 1976," 16, Institute of Texan Cultures Oral History Collection, OHT 721.9 F 711, UTSA Special Collections Digital Collections; "History," n.d., Folder "La Villita History Gen'l 1817–1940," SACS; "Art and Craft Chief Named," *San Antonio Express*, September 14, 1939.

56. While excavating, NYA workers found wine bottles, a five-gallon Meyer jug with copper wiring, jewelry, utensils, coins, animal bones, and a twenty-eight-foot-deep well chiseled with the date "1897" that Maverick had restored for Juarez Plaza.

"Interview with Harding Black, 1978," 33, Institute of Texan Cultures Oral History Collection, OHT 738.2 B 627, UTSA Special Collections Digital Collections; "Artifacts Found in Well at La Villita," *San Antonio Express*, November 17, 1939; "Old Villita Well Found," *San Antonio Light*, October 13, 1939; "Old Well Yields Curios," *San Antonio Light*, November 17, 1939.

57. "La Villita Progress Report."

58. "Artifacts Found in Well at La Villita"; "Villita Street Project Shaping Up Now," *San Antonio Light*, November 22, 1939.

59. Accuracy of this account has been called into question. "The 1722 Question," March 1, 2022, Folder "La Villita History, Gen'l 1817–1940," SACS.

60. "Sketch of Villita to Be Reprinted," *San Antonio Light*, November 6, 1939; "La Villita History Project Approved," *San Antonio Light*, May 7, 1940.

61. Maverick received letters from the state administrators of fifteen states and Puerto Rico. Hubert Y. Atherich (NM) to Maury Maverick, February 20, 1940; Ernest P. Marshall (WY) to Maverick, February 27, 1940; S. Burns Weston (OH) to Maverick, March 2, 1940; Robert Wayne Burns (CA) to Maverick, February 28, 1940, Box 2L52, Folder "Subject Correspondence, La Villita: 1939–1941," MMC, DBCAH.

62. Maverick and Williams worked on the campus newspaper together. Mary Carolyn Hollers George, *O'Neil Ford, Architect* (College Station: Texas A&M University Press, 1992), 62.

63. In 1936, while director of the Works Project Division of the NYA, David Williams published a manual of design and standards that influenced the indigenous style and use of local building materials in NPS structures of the era. Muriel Quest McCarthy, *David R. Williams: Pioneer Architect* (Dallas: Southern Methodist University Press, 1984), 123, 133; David R. Williams and O'Neil Ford, "An Indigenous Architecture: Some Texas Colonial Houses," *Southwest Review* 14, no. 1 (1928): 60–74; Williams, "Toward a Southwestern Architecture," *Southwest Review* 16, no. 3 (April 1931): 301–13.

64. David Williams also helped Ford get his other major NYA job: Little Chapel in the Woods at Texas Woman's University in Denton, whose timeline overlapped with the La Villita restoration. David Reichard Williams, "Note 1: Some Reasons Why of All This Rambling Palaver," 1939, Folder "La Villita—Ordinance," SACS.

65. Shortly after arriving in San Antonio to work on the La Villita project, Ford met Wanda Graham, the daughter of Elizabeth Graham, custodian of the Spanish Governor's Palace. Wanda was involved in the La Villita restoration in her own right as a member of the mayor's committee on the project. Ford and Graham married during the summer of 1940.

66. Texas Writers' Project of the WPA, *Old Villita*, 3.

67. O'Neil Ford to J. C. Kellam, August 30, 1940, Folder "La Villita—Ordinance," SACS.

68. "La Villita Progress Report."

69. David Dillon, *The Architecture of O'Neil Ford* (Austin: University of Texas Press, 1999), 50–51.

70. Maverick repurposed old paving material from the street commissioner for streets at La Villita because they would "look like they've been there a hundred years." "Villita Beautified," *San Antonio Express*, November 4, 1939; "Native Plants Adorn Villita," *San Antonio Light*, March 16, 1941; "Maury Views Street Work with Steffler as Guide," *San Antonio Light*, December 22, 1939 (quote).

71. "Only Real Old Villita House Torn Down," *San Antonio Express*, June 6, 1940.

72. Upon seeing workmen knocking down a wall of the Cos House, Rena Maverick Green called the Water Board to immediately stop the work. Fisher, *Saving San Antonio*, 54.

73. WPA Arts and Crafts workers also created a mural of bas-relief tiles depicting the history of La Villita in the six-foot stone wall surrounding the house. Bess Carroll Woolfred to Mary Green, December 20, 1940; "VILLITA" (manuscript), 7, Box 4J200, Folder 4: "WPA Bexar Co.—San Antonio—La Villita," WPAR, DBCAH; Writers' Program of the Texas Works Projects Administration, *Along the San Antonio River* (San Antonio, TX: Clegg, 1941), 25, San Antonio Archival File, Villita, La, Brochure, SAPL.

74. "CCC Camp for Olmos Due to Open in April," *San Antonio Light*, August 6, 1939.

75. Lewis Fisher credits Maverick's cousin-in-law and Carnegie board member Nicholas Kelley with helping him obtain the grant, while Maverick's wife suggested it was his aunt's friendship with the Carnegies. Scarbrough, "La Villita: Convergence of Life, Politics and Art," 5; "Carnegie Gift Given Villita," *San Antonio Light*, December 15, 1939; Fisher, *Saving San Antonio*, 215n94; "Interview with Terrell Dobbs Maverick Webb," 5.

76. Maury Maverick to J. C. Kellam, April 29, July 2, 1940, Box 2L52, Folder "Subject Correspondence, La Villita: 1939–1941," MMC, DBCAH; "$50,000 La Villita Building Starts Monday," *San Antonio Light*, July 4, 1940.

77. "La Villita Progress Report"; "CCC Camp for Olmos Due to Open in April."

78. Writers' Program of the Texas WPA, *Along the San Antonio River*.

79. "NYA to Hold Regional Meet," *San Antonio Express*, July 9, 1940; "Bolivar Site Work Slated," *San Antonio Light*, July 9, 1940.

80. "Interview with O'Neil Ford," 5.

81. Texas Writers' Project of the WPA, *Old Villita*, 4.

82. O'Neil Ford to J. C. Kellam, August 30, 1940. Folder "La Villita—Ordinance," SACS.

83. Don Politico, "Don Hopes They Will Return Band Concerts," *San Antonio Light*, August 11, 1939.

84. W. N. Beard, "Restoring La Villita, Spanish Village in Texas," *Mexia [TX] Weekly Herald*, June 13, 1941.

85. Rena Maverick Green, "Letters to the Editor," *San Antonio Light*, August 12, 1939. For more on Green's preservation efforts, see Fisher, "The Preservation of San Antonio's Built Environment," 212.

86. Hernández-Ehrisman, *Inventing the Fiesta City*, 12–13.
87. Fred Mosebach, "Oldtimers Recall La Villita Back 75 Years," *San Antonio Express*, December 31, 1939.
88. Texas Writers' Project of the WPA, *Old Villita*, 17–19.
89. Black history was largely ignored in 1930s and 1940s histories of La Villita. The old St. Philip's property was not part of the 1939 La Villita project, but it was purchased later by the city in 1947. Only one article published in the *San Antonio Express* included mention of the "negro congregation" and seminary that took over the old German church. Mosebach, "Oldtimers Recall La Villita Back 75 Years"; "History," n.d., Folder "La Villita History Gen'l 1817–1940," SACS.
90. Williams, "Note 1."
91. George, *O'Neil Ford, Architect*, 68.
92. Lou Block to Thomas Hibben, Memorandum, June 14, 1940, Box 6, Folder "Texas Information Reports," Entry NC-35 86, RG 119, NACP.
93. O'Neil Ford's firm Ford, Powell & Carson designed the Villita Assembly Hall on what was once Juarez Plaza in 1958–59 and oversaw a major renovation of La Villita in 1981. McCarthy, *David R. Williams*, xii; Dillon, *The Architecture of O'Neil Ford*, 53.
94. "Questions Answers," *Texas NYA Digest*, July 1941, 3, Box 132, Entry NC-35 330, RG 119, NACP.
95. The one-thousand-worker estimate is drawn from the NYA project applications (dated August 1, October 10, 1939; June 25, 1940; and March 25, 1941). The NYA classified Mexican Americans as "white" and did not further organize workers by ethnic group, so no statistics or reports indicate how many Mexican American youth were employed in San Antonio. However, the final report of NYA activity in Texas declared that "90% of the NYA program operating in San Antonio and south of San Antonio to the border of Texas was composed of Mexican youth." Given the project's emphasis on the city's Mexican American population, it is fair to assume that most NYA workers at La Villita belonged to that ethnic group. "La Villita Progress Report"; Project applications in Box 55, Folder "Texas 7c"; Box 115, Folder "Texas 14 276 San Antonio, Bexar Co. 308"; Box 6, Folder "Texas 36, Box 129, Projects Applications Case Files, 1935–1941," Records of the Project Planning and Control Section (Entry NC-35 258), RG 119, NACP; "Final Report: National Youth Administration for the State of Texas," July 1943, Box 6, Final Reports of 46 State NYA Offices to July 1943, Records Created During the Liquidation of the NYA (Entry NC-35 329), RG 119, NACP.
96. "Probers Given Keys to City," *San Antonio Light*, September 29, 1939; "Guides Ready for Tourists at Villita," *San Antonio Light*, February 1, 1941; "Table Depicts Cos Surrender," *San Antonio Light*, April 3, 1941; "Villita Gift to FDR Brings Thanks," *San Antonio Light*, February 25, 1941; "Mayor Takes La Villita Items to Roosevelt," *San Antonio Express*, February 26, 1941.
97. "The Villita Ordinance."
98. A copy of Lou Block's report, dated January 31, 1940, is enclosed in letter from Tom Hibben to J. C. Kellam, February 1, 1940, Box 2L52, Folder "Subject

NOTES TO PAGES 176–178 267

Correspondence, La Villita: 1939–1941," MMC, DBCAH; Lou Block to Thomas Hibben, Memorandum, June 14, 1940.

99. Lou Block, "Hand Skills in Crafts and Defense," *High School Journal* 24, no. 3 (March 1941): 111, 117.

100. Mary Vance Green was the daughter of Rena Maverick Green, founding member of the SACS. "Arts and Crafts Plans Made for Spanish Village," *San Antonio Express*, September 17, 1939; "Basement of Old Power Plant Will House Villita School," *San Antonio Express*, February 16, 1940; "City Sponsors Crafts School," *San Antonio Express*, February 21, 1940; "Villita Will Be Hand-Made," *San Antonio Light*, December 8, 1939; "Project Will Find It's in Red," *San Antonio Light*, May 31, 1940; "Camacho's Kin Has Villita Job," *San Antonio Light*, March 13, 1941.

101. George Biddle's residence began in February 1940. In his final report, Biddle recommended the city ask the Mexican government to send a muralist to decorate La Villita; in addition, he advised exhibiting collections from South America, purchasing an art library, and converting a wing of La Villita into a museum of Mexican folk art. Afterward, Biddle began to turn the forty-odd sketches he made of La Villita into paintings with motifs of "Mexican slums," longhorn cattle, and produce markets with brightly colored wagons. "Artist Is Sought for La Villita," *San Antonio Light*, December 13, 1939; "City Approves Ad Contract," *San Antonio Express*, December 14, 1939; "Biddle Accepts as La Villita Artist," *San Antonio Light*, December 27, 1939; "Art Authority for City Here Soon," *San Antonio Express*, February 1, 1940; "Report Made on La Villita," *San Antonio Express*, February 1, 1940; "La Villita Artist to Depict Texas," *San Antonio Light*, March 6, 1940; "Art Future for S.A. Proposed by Artist," *San Antonio Light*, March 12, 1940.

102. Ethel Wilson Harris's work promoted traditional Mexican crafts as well as romanticized images of Mexican folk scenes and the "Chili Queens." Harris served as the first park manager at Mission San José in the 1940s and lived on the grounds for more than twenty years. "Interview with Ethel Harris, 1976," 47, Institute of Texan Cultures Oral History Collection, OHT 721.9 H313, UTSA Special Collections Digital Collections; Joel D. Kitchens, "Making Historical Memory: Women's Leadership in the Preservation of San Antonio's Mission," *Southwestern Historical Quarterly* 121, no. 2 (October 2017): 191–92; Fisher, *Saving San Antonio*, 162.

103. Head of ceramics Harding Black recalled Maverick's almost daily presence at La Villita. "Interview with Harding Black," 13, 20; "La Villita Reconstruction Project— Transcript of an Interview of Mary Vance Green," October 2, 1978, p. 8., Folder "Mary Vance Green," SACS.

104. Maury Maverick to Fenner Roth, December 2, 1939; October 1, 10, 12, 1940; Maverick to J. C. Kellam, August 29, 1940; Maverick to Aubrey Williams, February 3, 1941, Box 2L52, Folder "Subject Correspondence, La Villita: 1939–1941," MMC, DBCAH.

105. Maury Maverick to Jesse Kellam, November 20, 1939; Kellam to Maverick, Cross Reference Sheet, September 24, 1940 (telegram dated September 12, 1940), Box 2L52, Folder "Subject Correspondence, La Villita: 1939–1941," MMC, DBCAH.

106. Jesse Kellam to Maury Maverick, December 22, 1939, Box 2L52, Folder "Subject Correspondence, La Villita: 1939–1941," MMC, DBCAH.

107. Fenner Roth, oral history interview 2 (II), October 11, 1978, by Michael L. Gillette, p. 17, LBJ Library Oral Histories, LBJ Presidential Library.

108. Maury Maverick to Aubrey Williams, January 7, 1941, Box 2L52, Folder "Subject Correspondence, La Villita: 1939–1941," MMC, DBCAH.

109. Aubrey Williams to Maury Maverick, February 7, 1941, Box 2L52, Folder "Subject Correspondence, La Villita: 1939–1941," MMC, DBCAH.

110. "National Administrator Visits Texas Projects," *Texas NYA Digest*, August 1941, 1, Box 132, Entry NC-35 330, RG 119, NACP.

111. "Villita Plan Is Approved," *San Antonio Light*, September 14, 1939.

112. "An Address by Maverick, before the Regular Meeting of the City Commission of the City of San Antonio," March 14, 1940, 16–18, Box 2L61, Folder "Maverick (Maury, Sr.) Papers, Speeches: Jan.–Apr. 1940," MMC, DBCAH.

113. By March 1941, the WPA had completed seventeen thousand feet of new walkways, thirty-one stairways, and three dams, and planted over four thousand trees and shrubs along the riverbanks. "River Project," *San Antonio Light*, October 29, 1938; "Think," *San Antonio Express*, November 4, 1938; John Hutton, "Elusive Balance: Landscape, Architecture, and the Social Matrix," in Miller, ed., *On the Border*, 227.

114. "Think," *San Antonio Express*, March 11, 1939.

115. "Think," *San Antonio Express*, November 4, 1938.

116. "A Speech Delivered by Robert H. H. Hugman to the San Antonio Historical Association on April 19, 1975," pp. 4, 6, OHT 976.4351 H894, Oral History Collection, UTSA Special Collections Digital Collections.

117. "Interview with Robert H. H. Hugman, 1977," 13, Institute of Texan Cultures Oral History Collection, OHT 976.4351 H894, UTSA Special Collections Digital Collections.

118. Hugman recalled that "it was the women who . . . pushed my proposals." Amanda Cartwright Taylor, the president of the SACS, introduced Hugman in the chamber meeting to discuss his plan for the river. "Interview with Robert H. H. Hugman," 17–18; "A Speech Delivered by Robert H. H. Hugman," 22.

119. For accounts of the preservation efforts of Adina Da Zavala, Clara Driscoll, and the Daughters of the Republic of Texas over the "Second Battle of the Alamo" in the early twentieth century, see Hernández-Ehrisman, *Inventing the Fiesta City*, 90–93; Fisher, *Saving San Antonio*, 40–60; and Gregg Cantrell, "The Bones of Stephen F. Austin: History and Memory in Progressive-Era Texas," in *Lone Star Pasts: Memory and History in Texas*, ed. Gregg Cantrell and Elizabeth Hayes Turner (College Station; Texas A&M University Press, 2007), 57–63. For discussion of the mythologizing of the Alamo, see Holly Beachley Brear, *Inherit the Alamo: Myth and Ritual at an American Shrine* (Austin: University of Texas Press, 1995).

120. "Church Eyed as Monument," *San Antonio Express*, September 24, 1940; "Mission Property Deeded," *San Antonio Light*, November 17, 1940.

121. Charles Hosmer continued, "When one compares the writings of Mayor Maverick

and the statements of the founders of the Conservation Society, it is clear that Maverick was in the mainstream of San Antonio thinking—long on inspiration and short on research, with a highly developed sense of historical reality but with little appreciation of the need for professional assistance." Hosmer's assessment is not entirely fair nor accurate; while both Maverick and SACS members promoted a fantasy Spanish historical landscape, Maverick intentionally courted the opinions of professional experts in architecture and craft design, as evident with the La Villita project. Hosmer, *Preservation Comes of Age*, 1:289.

122. "Interview with Mrs. Ethel Wilson Harris," interviewed by Esther Mac Millan, July 24, 1971, p. 21, 906 Har/A-153, SACS.
123. "Maverick Given Praise on Village," *San Antonio Light*, August 9, 1939.
124. Marci R. McMahon, *Domestic Negotiations: Gender, Nation, and Self-Fashioning in US Mexicans and Chicana Literature and Art* (New Brunswick, NJ: Rutgers University Press, 2013), 38.
125. Fisher, *Saving San Antonio*, 202.
126. "La Villita Pictured to Business Men as Attraction for Tourists," *San Antonio Express*, January 18, 1940.
127. "Ad Fund Will Fete Visitors," *San Antonio Light*, January 18, 1940.
128. Maverick also wanted the Chamber of Commerce to move its publicity bureau to the Cos House at La Villita. "Mayor Opens S.A. Publicity Campaign," *San Antonio Light*, March 18, 1940; "65 Signs to Show Way to La Villita," *San Antonio Express*, October 10, 1940.
129. "Maury to Conduct La Villita Tour," *San Antonio Light*, January 22, 1941; "La Villita Gets Giant Oak," *San Antonio Light*, January 26, 1941; "Villita Shown at Open House," *San Antonio Express*, January 27, 1941.
130. Texas Writers' Project, *San Antonio: An Authoritative Guide* (San Antonio, TX: Clegg, 1938); Writers' Program of the Texas WPA, *Along the San Antonio River*. For a deeper discussion of the Texas Writers' Project and the American Guide Series, as well as a broader analysis of WPA programs in the state, see Ronald E. Goodwin, *The New Deal and Texas History: Saving the Past through Hardship and Turmoil* (Lanham, MD: Lexington Books, 2021).
131. Alex N. Murphree, "Williamsburg in a Sombrero: The Story of 'La Villita' in San Antonio," February 1940, Folder "La Villita—Restoration—1939," SACS.
132. "First of School Groups Visit La Villita Project" *San Antonio Express*, December 12, 1939; "First Play Enters Contest," *San Antonio Express*, December 31, 1939; "Students Visit Villita Project," *San Antonio Express*, January 13, 1940; "Texas U. Students Inspect La Villita," *San Antonio Light*, July 9, 1940.
133. "First Class Visits Villita Project," *San Antonio Light*, December 12, 1939.
134. "40 Arrive Here to Make Mexico Tour," *San Antonio Express*, March 31, 1940.
135. "Camacho Aide Visits Mayor; Sees Village," *San Antonio Express*, December 1, 1939; "Camacho Kin to Visit S.A.," *San Antonio Light*, May 8, 1941; "2nd Division Review by Visitors," *San Antonio Express*, October 8, 1940; "Latin-American Officials Hail Reception by Army," *San Antonio Express*, October 9, 1940.

136. "La Villita Given Fountain, Statue," *San Antonio Express*, October 25, 1940; "Tall Statue Given Villita," *San Antonio Light*, December 9, 1940.

137. "Msgr. Schnetzer to Bless Statue at La Villita on Guadalupe Day," *San Antonio Express*, December 11, 1940; "Yule Pageant to Be Given at La Villita," *San Antonio Express*, December 21, 1940; "Christmas Programs Held at La Villita," *San Antonio Express*, December 25, 1940; "Miracle Play Set Tonight," *San Antonio Express*, December 30, 1940; "Latin-American Day Slated at La Villita," *San Antonio Express*, February 12, 1941; "Entertainment Planned for New Citizens," *San Antonio Express*, April 13, 1941; "City Sets Sunday as Civic Day," *San Antonio Express*, April 1, 1941; "City to Show Wares Today, Army Monday," *San Antonio Express*, April 6, 1941; "Many Visit City Buildings on Civic Day," *San Antonio Express*, April 7, 1941.

138. The river festival moved permanently from San José Mission to La Villita in 1947 and took on its current name, "A Night in Old San Antonio," in 1948. "Festival at River Planned," *San Antonio Express*, September 17, 1940; "Conservation Society Plans River Jubilee," *San Antonio Express*, October 6, 1940; "River Jubilee Set for Friday Night," *San Antonio Light*, October 13, 1940; "La Villita Given Fountain, Statue"; Fisher, *Saving San Antonio*, 207.

139. "King to Enter Fiesta by Water," *San Antonio Light*, March 26, 1941; "Villita Gallery to Have Exhibit," *San Antonio Express*, April 18, 1941; "It's Official! Mayor Proclaims Fiesta Week," *San Antonio Express*, April 19, 1941; "San Antonio Ready for Fiesta Week," *San Antonio Express*, April 20, 1941.

140. Mary Ann Villareal, "Becoming San Antonio's Own: Reinventing 'Rosita,'" *Journal of Women's History* 20, no. 2 (Summer 2008): 86. Laura R. Barraclough describes how the small population of *ricos*, the wealthy Mexican exiles who fled to San Antonio during the Mexican Revolution, participated in the promotion of Mexican cultural heritage through elite forms like the Spanish-language newspaper *La Prensa* and *charrería*, a sport based on traditional equestrian practices. Barraclough, *Charros*, 70.

141. Daniel D. Arreola, "Urban Ethnic Landscape Identity," *Geographical Review* 85, no. 4 (October 1995): 523.

142. "Mayor Host to TVA Chief on Visit Here," *San Antonio Express*, April 12, 1940; Fenner Roth, oral history interview, 17.

143. Fenner Roth to Floyd McGowan, January 10, 1940, Box 2L52, Folder "Subject Correspondence, La Villita: 1939–1941," MMC, DBCAH.

144. State administrator Jesse Kellam visited La Villita at Christmas time to assist in the filming of a scene and cohost a Christmas party for the NYA workers. Jimmie Lederer to J. C. Kellam, November 7, 1939, Box 6, Folder "Texas Information Reports," Entry NC-35 86, RG 119, NACP.

145. "Film to Record NYA Activities," *San Antonio Express*, October 14, 1939; "NYA Youths Guests of City at La Villita," *San Antonio Express*, December 24, 1940; Letters between Maverick and NYA officials in Box 2L52, Folder "Subject Correspondence, La Villita: 1939–1941," MMC, DBCAH.

146. Henderson, *Maury Maverick*, 200.

147. P. J. Kelly, "Letters to the Editor—A Tourist's Reaction to La Villita," *San Antonio Light*, February 21, 1941.

148. "S.A. Asks Where but Okehs Villita Project," *San Antonio Light*, February 5, 1941.

149. In fiscal year 1939, which ran from May 31, 1939, to May 31, 1940, La Villita cost the city almost $17,500: $11,000 was spent on purchasing land, $3,312.94 on buildings, $800 on labor operations, $750 on salaries, $500 on office supplies, $50 on telephone and telegraph operations, and $50 on miscellaneous supplies. Meanwhile, $837.06 was spent on "other expenses" and $175 for "other sundry charges." "Costs of Villita Shown in Budget," *San Antonio Light*, March 15, 1940.

150. "Interview with Louis Lipscomb," 11.

151. "Over Your Coffee Cup," *San Antonio Express*, April 15, 1940.

152. The city of San Antonio issued a special three-cent tax in 1939 for the purpose of "advertising the city nationally." The city council spent almost $80 from this municipal advertising fund on celluloid for NYA film, since it would "give the city national publicity," and another $375 for the printing of 200 copies of *Old Villita*. "Village Project Taps Ad Fund," *San Antonio Express*, October 13, 1939; "Villita Film Made for NYA," *San Antonio Light*, October 18, 1939; Don Politico, "Maverick's Year in Office Tumultuous," *San Antonio Light*, June 2, 1940.

153. According to Fenner Roth, the NYA district director, C. K. Quin made "La Villita a campaign issue about how money was being wasted" and said that "if elected he was going to tear it down." Fenner Roth, oral history interview, 31.

154. "Mayor Praises Past Work," *San Antonio Express*, May 1, 1941.

155. "Mayor Denies Mexicans to Vote Solid," *San Antonio Light*, May 2, 1941; "Mayor Says He Improved City's Health," *San Antonio Light*, May 4, 1941; "Mayor Claims Mexicans Helped," *San Antonio Light*, May 12, 1941.

156. C. K. Quin had used this strategy of calling Maverick "red" in the 1934 House campaign as well. "Money Waste by Mayor Charged," *San Antonio Light*, May 5, 1941; Cunningham, *Bootstrap Liberalism*, 80.

157. Fairbanks, "Public Housing for the City as a Whole," 404.

158. In the initial election, Quin received a plurality with 17,435 votes to Maverick's 16,202. Doyle, "Maury Maverick and Racial Politics in San Antonio," 217–20; "Quin Silent on Appointees as Speculation Increases," *San Antonio Express*, May 29, 1941.

159. Doyle, "Maury Maverick and Racial Politics in San Antonio," 217–19.

160. "Mayor Dedicates Child, La Villita," *San Antonio Light*, June 1, 1941.

161. The park commissioner Henry F. Hein "long believed that La Villita, being a public property, belonged in his department." "Mayor Gives Up La Villita," *San Antonio Light*, May 29, 1941; "Quin Inspects Maverick Pet," *San Antonio Light*, June 25, 1941; "Villita Gets Quin Approval," *San Antonio Light*, June 26, 1941.

162. "Villita Gets Quin Approval"; "Villita to Lose Hidalgo Statue," *San Antonio Light*, June 25, 1941.

163. In early 1941, members approved a motion to create a Board of Governors "with power similar to the Carnegie Library board" and to "remove La Villita from political control to a great extent." Board of Trustees and Stockholders of La

Villita Corporation meeting minutes, January 17, 1941, p. 25, Box 3J119, Folder 3: "Official Minutes April, 1940–Dec., 1952," La Villita Corporation Records, 1939–66 (hereafter LVCR), DBCAH; "'La Villita' Incorporated by Maverick," *San Antonio Express*, April 19, 1940.

164. Board of Trustees and Board of Governors meeting minutes, March 14, 1941, 67, Box 3J119, Folder 3, LVCR, DBCAH.

165. Board of Trustees and Board of Governors meeting minutes, December 10, 1941, 105, Box 3J119, Folder 3, LVCR, DBCAH; "National NYA Administrator Praises La Villita Project," *San Antonio Express*, August 23, 1941.

166. "Maverick Dream Becomes a Reality," *San Antonio Light*, February 6, 1942; Board of Governors and Board of Trustees meeting minutes, May 9, 1942, 112, Box 3J119, Folder 3, LVCR, DBCAH.

167. "NYA Speeds Defense Work," *San Antonio Express*, August 10, 1939; "Maverick to Stay Out of House Race," *San Antonio Light*, June 16, 1940 (quote); "Villita Unit to be Started," *San Antonio Light*, July 4, 1940; "$50,000 La Villita Building Starts Monday."

168. Board of Trustees and Board of Governors meeting minutes, March 14, 1941, 69, Box 3J119, Folder 3, LVCR, DBCAH; "Bexar_La Villita Under the Stars_captions."

169. The Red Cross began occupying buildings on January 1, 1942. Board of Trustees and Board of Governors meeting minutes, December 10, 1941; May 9, 1942, 104, 111–12, Box 3J119, Folder 3, LVCR, DBCAH; "Red Cross Plans War Relief Drive," *San Antonio Express*, December 11, 1941; "Bolivar Bldg. to Red Cross," *San Antonio Light*, December 31, 1941; "27 Women Take Red Cross Class For Motor Corps," *San Antonio Express*, January 19, 1942; "39 Classes Held for Home Nursing," *San Antonio Express*, March 12, 1942.

170. Maverick F. Fisher, "Maury Maverick, La Villita, and the Pan-American Dream" (Honors Program University of Texas at Austin, May 13, 1998), 21–22, Call No. 921 MAV, SACS.

171. Eight groups of newspaper reporters and editors representing Brazil, Chile, Colombia, Costa Rica, Cuba, Dominican Republic, Ecuador, El Salvador, Guatemala, Honduras, Nicaragua, Panama, Paraguay, and Venezuela came through La Villita. Board of Trustees and Board of Governors meeting minutes, April 12, September 1, 1943; May 9, 1944, 117, 123, 125, Box 3J119, Folder 3, LVCR, DBCAH; Photographs (captioned), Box 3X9a (Photographs), Folder "La Villita Board, Pan American Rep./South American Rep.," LVCR, DBCAH.

172. In 1944, for example, $6,391.81 was collected in rent from organizations and groups that brought in 62,044 visitors, up from $2,346.00 and 34,990 visitors in 1942. "Price Increase Is Asked at La Villita," *San Antonio Light*, June 12, 1941; Joint meeting of the Board of Governors and Board of Trustees, February 28, March 16, 1943; May 9, 1944, 113, 115, 125; Meeting of the Board of Directors, May 9, 1945, 129, Box 3J119, Folder 3, LVCR, DBCAH.

173. Joint meeting of the Board of Governors and Board of Trustees, December 10, 1945, 138, Box 3J119, Folder 3, LVCR, DBCAH.

174. Board of Trustees and Board of Governors meeting minutes, September 1, 1943; May 9, 1944, 123, 126, Folder 3, Box 3J119, LVCR, DBCAH; "La Villita Physical Properties," n.d., Box 3J119, Folder "d. Financial Record," LVCR, DBCAH.

175. "Looking South for Yule," *San Antonio Light*, December 20, 1942.

176. "Interview with Lasca Fortassain, 1977," 12, Institute of Texan Cultures Oral History Collection, OHT 923.2 F736, UTSA Special Collections Digital Collections.

177. Board of Trustees and Board of Governors meeting minutes, April 12, September 1, 1943; May 9, 1944, 117, 122, 125, Box 3J119, Folder 3, LVCR, DBCAH.

178. La Villita Advisory Board and La Villita Corporation meeting minutes, June 18, 1952, 7, Box 3J119, Folder 3, LVCR, DBCAH.

179. La Villita Advisory Board and La Villita Corporation meeting minutes, June 18, 1952, 19, Box 3J119, Folder 3, LVCR, DBCAH.

180. "Final Report: National Youth Administration for the State of Texas."

181. Walter Prescott Nobb to Maury Maverick, February 5, 1941; Maverick to Nobb, February 11, 1941, Box 3J120, Folder "2. Maury Maverick manuscript notes by Charles Curtis Munz," LVCR, DBCAH.

182. "Mr. O'Neil Ford Interviewed by Charles Hosmer and Elsa Watson, June 29, 1971," 9, Folder "O'Neil Ford 725.9/A-158," SACS.

Conclusion

1. Woodward, "Allies in Aims," 1078.

2. Hayden, *The Power of Place*, 1.

3. Bruggeman, "Introduction," in *Commemoration*, ed. Bruggeman," 6.

4. Thompson M. Mayes, chief legal officer for the National Trust for Historic Preservation, makes the case that people care about historic places for diverse reasons ranging from their beauty and sacred qualities to their role in identity and community building, to a sense of continuity with the past. Mayes, *Why Old Places Matter: How Historic Places Affect Our Identity and Well-Being* (New York: Rowman and Littlefield, 2018).

Index

Stephanie Gray holds a PhD in US history and an MA in public history from the University of South Carolina, and a BA in history from Mount Holyoke College. She is assistant professor of Public History at Duquesne University in Pittsburgh, PA, where she won the McAnulty College Pre-Tenure Faculty Award for Excellence in Teaching.